EVOLUTION: A CRITICAL APPROACH

David Lowsley

ATHENA PRESS
LONDON

EVOLUTION:
A CRITICAL APPROACH
Copyright © David Lowsley 2005

All Rights Reserved

No part of this book may be reproduced in any form
by photocopying or by any electronic or mechanical means,
including information storage and retrieval systems,
without permission in writing from both the copyright
owner and the publisher of this book.

ISBN 1 84401 413 4

First Published 2005 by
ATHENA PRESS
Queen's House, 2 Holly Road
Twickenham TW1 4EG
United Kingdom

Printed for Athena Press

Contents

Chapter I
INTRODUCTION — 7

Chapter II
AND IN THE BEGINNING… — 9

Chapter III
THE DARWINIAN THEORY — 13

Chapter IV
THE ORIGIN OF LIFE — 29

Chapter V
NATURAL SELECTION, VARIATION AND MUTATION — 48

Chapter VI
LAMARCK: A THEORY OF EVOLUTION — 56

Chapter VII
THE EVOLUTION OF THE MAIN CLASSES — 63

Chapter VIII
PROBLEMS? — 70

Chapter IX
CAN DARWINISM SURVIVE? — 82

Chapter X
PURPOSE? — 90

Epilogue — 96

Bibliography — 104

EVOLUTION:
A CRITICAL APPROACH

Chapter I
INTRODUCTION

My fascination with the natural world started when as a child, and as a consequence of the outbreak of the Second World War, I was evacuated into the country to escape the bombing. My next four years were spent living on a farm and experiencing at first hand nature in all its various aspects. It was an exciting time for a child. In what was to me a strange and fascinating world I wanted to know the answers to so many questions. Why did the lambs always appear in the spring? Why were certain crops harvested at different times of the year? Why did some birds choose to nest on the ground while others nested at tops of trees? The answers I received from the two young ladies who ran the very overcrowded village school did not always seem to make much sense. Religion played a very important part in village life, and it was supported by a church and a chapel. The latter was to satisfy the needs of the farm labourers and their families, and the church provided a similar function for the farmers and tradespeople. There was also a Sunday school for the children. Not having any established place in the community it was deemed fit that I should attend the church service in the morning, the Sunday school in the afternoon, and the chapel at night.

So my relationship with this strange world was also linked very much with its religious aspect. To be honest I did not find this was any great help in providing believable answers to my questions. Much later in life, I was to realise how religion and ignorance were inexplicably linked. Religion provided answers which either science could not supply, or which could only be understood by education. These early experiences shaped my future, and I have been asking questions and attempting to find the answers with various degrees of success ever since.

Subsequently I was to spend many years teaching biology and

have been saddened by the fall off of questions from pupils as they became older. At student level there was a blind acceptance of what I taught, and what I had to teach was based upon the contents of the set textbooks which were occasionally brought up to date by incorporating any new discoveries into their texts. There was never any apology in the books to say sorry, in view of what we now know the information we supplied in the past was wrong. The students absorbed and regurgitated the facts and on that basis were qualified to either teach or study at a higher level.

During this time I of course taught the theory of evolution, which was based upon modified Darwinism. No student ever questioned the validity of what I taught. This was not teaching, it was as much indoctrination as was the teaching of religion. It called for blind faith, and that is what it received. When I asked a bright student if he had considered research as a career his reply was, 'Well, there isn't much left to find out.' Perhaps there is a case for teaching what we don't know rather than what we think we do know. This however would be difficult to achieve unless our present understanding is subjected to questioning and unbiased scrutiny. Our present explanation of evolution has always worried me. It is based upon too many as yet unanswered questions. I have therefore attempted to pose some of these questions and at the same time examine the validity of the answers which present-day science produces.

Chapter II
AND IN THE BEGINNING…?

The human brain provides us with both the ability and the desire to question the purpose for our existence. The reason for this is not difficult to understand, because just about everything on Planet Earth appears to have a purposeful function. This fact is emphasised when we compare it with the other planets in our solar system. The earth could have been tailor-made to support the diversity of life which was required in order for us to evolve. In doing so it evolved the human brain, which has advanced far above that required by the earth's other life forms to assist them in their survival. It is our brains, and their products, which appear to have been the destiny of evolution, and everything else, including our bodies, are merely cast in supporting roles. This of course raises the question of how life evolved and became so diversified on a planet that also appears to have evolved in order for it to become ideally suited to support it. The different life forms have become integrated with the function and needs of all other life forms. Raw materials required for the production and maintenance of life use other life forms to recycle the materials. Plants supply the oxygen required for animal life, which in turn provides the carbon dioxide necessary for the plants to flourish. Carbon is thus recycled, as are the mineral requirements needed by each new generation of life. The study of ecology is the study of the various cycles, the carbon cycle, the nitrogen cycle, the water cycle, and more importantly the understanding of the balance of nature. The whole process of life is of course dependent upon the ability of plants through the process of photosynthesis to harness the sun's energy. Without it there would be no life and no evolution. The whole system is of ordered precision, which we are expected to believe occurred as a result of a series of chance occurrences. Is it any wonder that it calls for the concept of a creator to explain its existence? Religions arise from this concept, but in themselves provide no tangible proof to account for our existence. The theory of evolution arose in an attempt to provide acceptable proof to explain how man has evolved, and at its base is the supposition that it is the result of a series of chance events any one of which could have changed the whole course of evolution.

Life, we are led to believe, started through the chance ability of inorganic materials to assemble themselves into self-replicating molecules, which over a vast period of time evolved by natural selection into man. Scientists generally agree that the odds of life having started in this way and having been capable of reproducing itself is mathematically impossible to calculate, but since we are here then it must have occurred. Despite this, the search is on to find other intelligent life forms in the universe who will have evolved the technology and the ability to make contact with us. What would this prove? It is of course impossible to prove that life does not exist elsewhere, and all that can be hoped for is that life forms can be found, and that the earth is not unique in the universe.

I contend that the theory that life has supposedly evolved by natural selection resulting from chance changes being retained if they provide beneficial factors, and being lost by the survival of the fittest concept if they don't, may well be true. It alone does not, however, account for the way evolution has occurred. Nor does it explain many factors which have been ignored in the present theory of evolution. It is my intention to consider these in some detail. Darwinism does however provide a basis upon which one can study the origin and development of life. It was never intended to be anything more. Scientific research inevitably produces more questions than answers. In the case of Darwinian evolution, it has been easier to allow it to provide all the answers, and when this proved difficult, then to cease asking the questions.

Let us consider some of the issues. From the assumed chance occurrence of a self-replicating molecule life then proceeded to evolve with no predesigned plan to produce us. In order for this to happen two things had to occur, neither of which fits into the concept of the Darwinian theory. First of all reproduction, especially sexual reproduction, became the driving force and purpose of all living things. Why, I ask, should this be so? Competition with other life forms, yes, but why reproduction? One can argue that if life did not reproduce then evolution would not occur. This is of course true, but this would also suggest that the purpose of the original self-replicating molecule was to evolve. The second thing which had to occur was among living organisms' deaths. Natural death is accepted as normal after they have reproduced. Their actual lifespan is generally dependent upon the time taken to mature from birth so as to be physically capable of reproducing, and to allow sufficient time to enable the next generation to be capable of surviving. This is not advantageous to the individual concerned, but is only for the benefit of the next generation, and of course for evolution to take place. Once again this points to purpose not chance.

I stated earlier that there were two theories to explain the reason for our existence, one the creative theory and the other the evolutionary theory. I contend that a third theory begs to be considered. At the present point in time man has evolved from all the life forms that exist and have ever existed on earth. Their function and purpose were to ensure that we would eventually emerge, and with us the human brain. Reproduction results in a seventy-year turnover of millions of brains with the inevitable production of those which are capable of building upon the Newtons and Einsteins of the past in attempting to understand the whole universe, its origin, function and future. Thus although attempts are made now to understand concepts which at the present time we are incapable of understanding, we are nevertheless on the threshold of building machines which will evolve and take over the role of the more highly developed human brains without suffering the handicap of survival faced by all living organisms. Let us suppose that this was the original goal, and that the first self-replicating organisms did not arise by chance, but were placed on earth to fulfil their function to produce machines which would in the distant future be able to enter a new territory of understanding and knowledge. We then are merely the tools to ensure that this takes place. I have stated that self-replicating molecules were the origin of life. These were programmed for different roles, some would remain as simple unicellular organisms, others evolve into photosynthesising plants, and others with the role of evolving into advanced animal life forms. It is rather like the eleven players at the start of a football match. It is not until the game has started that each player's role can be defined. This supposition may sound far-fetched, and I would be the first to admit may well be wrong, but it does seem more plausible than the concept that all life started from one type of self-replicating organism.

Before one scoffs at the concept that life is programmed to replace human brains with machines, one should consider where we will be in say a thousand years' time, or even a million years, which is only a fraction of the time the dinosaurs ruled. Time is of course immaterial since it has no beginning or end, and therefore it plays no part in any consideration of how long it takes. The third theory reminds one of the child painting a picture of God. The teacher said, 'But no one knows what God looks like.' 'They will when I finish,' said the child. Perhaps when evolution has finished then, and only then, will our place in the universe be explained.

Let us therefore now look at why the apparently purposeless role of man has been so readily accepted, and how science has ignored certain facts in order to support the aimless concept of Darwinian evolution.

Chapter III
THE DARWINIAN THEORY

There can be few people today who have not heard of the name Charles Darwin and its association with evolution. This is hardly surprising. His theory, or to be more accurate his hypothesis of evolution, has laid the foundation for our understanding, often erroneous, of the science of biology. More than that, it explains with seemingly reasonable logic and clarity the answer to the question that has puzzled people from the very beginning. What is the origin of all the vast variety of living things, all with differing degrees of complexity, and ending, so it seems, with man at the top?

Religions of course have provided, and still do provide, an answer to these questions, but they rely upon blind faith with little hard evidence of support. Darwin seemingly supplied the answer with the words which have become incorporated into everyday language – the survival of the fittest. To be fair, Darwin never actually used this phrase. It was in fact coined by Herbert Spencer, who had reflected upon the concept of evolution before Darwin published his theory. As a firm believer in Lamarckism, which I will discuss later, he was never taken seriously. Nevertheless the concept was easy to grasp. It seemed perfectly logical, and it made sense that the strongest must always survive over the weak.

After the publication of *The Origin of Species* the concept became a very acceptable proposition. It supported the class structure, and validated the domination of one race over another. Here was science apparently demonstrating that the true destiny of man not only could, but should be achieved by the strong surviving and the weak being removed. The door had now been opened to any dominant political group using this concept to justify the domination of its own doctrine. The Third Reich in

Germany made use of this distortion, with the horrific results of which we are all aware. Darwin's proposition had far greater impact and consequences than he had ever intended. The debate that followed the publication of *The Origin of Species* was based upon the hypothesis that man had evolved from the apes. It won the day, not because of any scientific evidence, but because it reinforced the belief that the white races were obviously more highly evolved than the less civilised black natives. Domination by white supremacy was therefore justified, because science had endorsed that this was the natural order of things. It is not difficult therefore to understand how such an acceptable interpretation of Darwin's theory did so much to establish the broader concept of evolution. It was this, and its acceptance of how life has evolved, that has done much to reduce the need for further examination of the evidence.

The rest of Darwin's theory was just as satisfying. Simple life forms must have evolved before more complex forms. Machines, buildings etc. all follow this principle. Man could not have built a cathedral before he was able to build a house. Well, that basically was it. To the average person nothing else was required, and even today that would about sum up most people's knowledge of an idea that is fundamental to our understanding of the natural world.

Who was this man who has had such an impact upon our understanding of the living world? He was born in 1809 in an upper-class background, and went on to study medicine. He changed his idea of becoming a doctor and enrolled on a course of study with a view to becoming a clergyman. While at Cambridge he studied botany and this, coupled with a passion for natural history, led to him being recommended for the post of ship's naturalist aboard *HMS Beagle*. The ship's mission was to chart stretches of the South American coastline. As the ship worked its way around the continent, Darwin, who appears to have spent much of his time ashore, observed and collected specimens of all kinds of plants and animals. Upon his return to England he became a respected person in the scientific community, working on subjects as diverse as the production of a comprehensive monograph on barnacles, a study of trophism in plants, and the role of earthworms in soil fertility. The point being made that even without *The Origin of Species*, Darwin would still have been regarded as a noted and respected Victorian scientist. He however assured his place in history by formulating his theory of evolution, which he was extremely reluctant to publish. Was this because he anticipated the reaction it would create? Well, certainly not from the scientific community. Perhaps he worried what its effect would be on the church, or was it because he felt that there were flaws in the theory which would be subjected to objective criticism?

Matters were however taken out of his hands; Darwin received a letter from the naturalist Alfred Wallace, who worked in the West Indies. He too had developed a theory of natural selection, and asked Darwin to review his paper. This had a tremendous impact upon Darwin. He saw his work on the theory of evolution being snatched away. Wallace's paper was published along with an essay from Darwin. A year later Darwin published *The Origin of Species*. His work was so extensive and thorough that he is of course known as the main author, and even Wallace agreed that Darwin deserved the full credit.

Darwin had of course backed up his supposition in great detail, but how widely his book was read and fully understood is hard to judge. Its initial impact was more concerned with the downgrading of man from divine creation to an offshoot of the apes. I would suspect that few people today have bothered to read it. Certainly I doubt if it is to be found in any classroom as a working textbook. Equally I doubt if there is any biological textbook that does not refer to Darwinism in some way. The result is that by this method, all the major problems of evolution are solved. With the new discoveries in genetics and biochemistry, Darwinism can be brought up to date without affecting the general acceptance of his theory. Let us now look at the ways in which Darwin arrived at his conclusions.

Darwin, having travelled the world aboard the *Beagle*, and having discovered an even greater variety of living organisms than even he had been aware existed, formulated his hypothesis to explain this great diversity. He noted that although species tend to reproduce in greater numbers than can survive, the actual population generally remains constant. The struggle for survival therefore results in some surviving and some dying. Since the population produces variation, in other words they are not all the same, then some would have an advantage over others, hence the survival of the fittest. It seemed obvious that those variations which proved to be advantageous would be passed on to the next generation by natural selection. Any of course that were not advantageous would die out. This would lead over a period of time to species changing and becoming better suited to their environment. Since not all variation would be the same for the whole of the population, and since the environment, and the survival needs of the species, would not necessarily always be constant factors, this would allow some members of the species to change, and so adapt to different conditions such as changed habitats, or different foods. In this way, rather than be in competition with the rest of the population they would be able to branch out and produce a new population which would eventually evolve into a new species.

Darwin backed his ideas with evidence from the breeding of domestic animals. Here, by artificially selecting and cross-breeding, different strains can be produced with specific characteristics, as for example in the domestic dog. It is probably worth pointing out that this method, although producing a vast array of different breeds, has not resulted in any new species, just many variations of dogs. This is an accepted criticism and is countered by saying that if the variations become separated geographically then the differences will continue to occur and increase, so as to become so exaggerated that it would lead to a new species which would be unable to breed with its original ancestors.

The problem here is that if two populations are in fact geographically separated then there is no way of knowing if they are capable of interbreeding. It is rather interesting that a recent report in the press and a subsequent television broadcast featured the surprised owner of a Great Dane bitch which had given birth to a litter of pups fathered by their pet Corgi. The physical variation in the differences between the two breeds had, it may be said, not proved insurmountable, and nature had taken its course. What we do know is that if two closely related species interbreed then the result is a hybrid which is usually sterile. An obvious example of this is the horse and the donkey interbreeding to produce a mule, which is sterile. The horse and the donkey have therefore remained as separate species. What is the advantage of this? On the face of it, it would seem a quick and efficient way, if the hybrids could breed, for the production of a new species. However by being unable to breed the individual species are protected, and this keeps the gene pool flowing in one direction through each species.

The difficulty is that although this could account for the changes which are rather small, it would not account for the major changes which would need to have taken place to separate major groups of both the invertebrates and vertebrates, as well as those in the plant kingdom. The differences are too great. Biologists now explain this on the basis of mutations arising which would then come under the same Darwinian principle of natural selection. Small changes result when errors are made in the replication of the genes and the passing on of these in sexual reproduction. Major changes could have occurred when the genetic material was affected by external factors, such as certain chemicals, radiation etc. which results in dramatic mutations.

Mutations do of course occur, and the argument is that although most of these are detrimental to survival, some provide the organism with certain advantages. These, like the naturally occurring variations, can be passed on to the next generation. Could this account for the major changes needed to produce the main groups of organisms? We are, I contend, now entering an area in which evolutionary theory becomes a little more difficult to accept. However, riding as it does on the back of Darwin's theory, it has become generally accepted without looking too closely at the supporting evidence. It is this evidence which needs to be examined. Here we are not just concerned with examining the variation of living organisms, but the very origin of life itself on this planet. Life must have evolved from non-living material, and therefore evolution, and the process which brought it about, must be applicable not only for the production of complex organisms, but in the beginning to the evolution of non-living materials into self-replicating organisms.

Geographical isolation has to be an essential factor if mutations are to be exploited to produce the changes which would prove significant in producing new species. In a large population these would simply be swallowed up, while in a small population they would have a greater chance to dominate. The difficulty here is that the mutation would have to be sufficient to produce the physical changes required in order to be responsible for the changes of a fin into a leg, or a leg into a wing. The problem is that the transitional stages would serve no purpose, and in the process of change would be of no use for either function. How this could be deemed an advantage over others where the mutation had not occurred is difficult to see.

Another important factor is that further mutations would also have to occur and in the correct sequence if the changes were to prove advantageous. It is no use fins becoming limbs if changes in body shape and the ability to breathe air fail to happen. These changes, if they are to be effective, cannot occur gradually, as this would be a disadvantage. A gradual reduction in the structure of a fin designed to be effective for swimming would give no advantage to the fish, which would in fact be at a disadvantage in the competition for survival. In the rule of survival of the fittest they would in fact die out. Should the change be achieved quickly, and the lack of intermediary stages in the fossil record would suggest that this is what in fact did occur, then the animal may have gained some advantage. This would however depend upon changes in its original habitat in order to be able to utilise these changes. The drying up of an aquatic habitat, for example, thus allowing the animal to survive. The same mutation would of course have to occur among other members of the species so that the mutation could be passed on by reproduction, otherwise the original mutation would be lost. The question which arises from this is, do mutations occur spontaneously or are they a response to environmental change? This may seem unlikely, but it is equally unlikely that they occur and spread through a population on the off-chance that they may at some time in the future prove advantageous for survival.

The confusion in Darwinian evolution is the difference between variation and the production of species. Let us return to the survival of the fittest, in which, by definition, more offspring are produced than can survive. Under these conditions small variations resulting from the combination of the parents' genes not only occur, but can lead to the production of distinct and significant variations. This is obvious in the production of the different races in the human population. Would this however lead to a new species? It is difficult to see how.

Much is made in Darwinism of the famous Darwin finches. Darwin used the finches inhabiting the Galapagos Islands as observable evidence of evolution at work. He concluded that they had probably originated from a flock of finches which had arrived on the islands some time in the past and had evolved their differing features. This is much cited as an example of natural selection at work. Among the birds that Darwin collected were thirteen types of finch. Some were unique to individual islands, while others were distributed on one or two islands which were close together. He did not keep careful records of which birds came from where, which would suggest that the importance as evolutionary evidence which is given to the finches today, was of no particular importance to Darwin at that time. He was not in fact even sure that they were indeed separate species, and it was only when they were examined back in England that ornithologists confirmed that they were indeed different species. The most obvious difference was the shape of their beaks.

I think we have to assume that the flock of birds on their arrival at the islands did all have the same features. This is a supposition, although one which I am prepared to accept. The birds today are all greyish-brown and short-tailed, although in some the males are black. They resemble one another closely in courtship displays and in their nests and eggs. The striking difference is in their beaks, which have different shapes, and in their feeding habits. Each group has taken up different habitats on the islands, such as the coast, and forest areas. Those birds with pointed beaks are able to feed on insects in the cracks of tree bark, while those with thicker, stouter beaks use seeds and nuts as a food source.

Darwinists have therefore concluded that a new species will arise by the gradual accumulation of adaptations to different environments.

Here I contend that one has to be sure of how one applies the definition of species. Members of a species are by biological definition generally similar in appearance and distinct from members of other species, with no intermediate forms. Members are capable of interbreeding and producing fertile offspring. Does this then apply to the finches in question? Since the specimens were dead and preserved when examined and declared as different species, it is difficult to see how the criteria of interbreeding could be applied. One is therefore left with similarities of appearance and behaviour. Without wishing to be too cynical, had Darwin collected different members of the human race on his travels, and then these had been subjected to the same criteria to study the differences in head and body shape, skin colour, eating behaviour and diet, as well as habitats, he would have concluded that several human species had evolved. It would be a brave scientist who declared that the Australian aborigine, the African pigmy and the Arctic Inuits are all different species. Yet Darwinism is founded on the fact that new species arise by the gradual accumulation of features enabling adaptation to different environments.

Now, I have already accepted that individuals best suited to a particular environment have a higher probability of leaving offspring than those less well adapted. The problem with this is that it will lead to specialisation and therefore increase the risk of extinction. The finches have adapted to exploit certain foods because of the changes in their bills. Whether the change of diet occurred after the bills changed shape or because they had changed is a question I will return to later. The point is that once they had changed, and the birds had become specialised to exploit one food source, then they became dependent upon it for survival. To say however that the finches have evolved into different species is rather optimistic. At the best one would classify them as a sub-species. The changes are hardly significant in the explanation of evolution. A beak is still a beak and a bird is still a bird. Were one to examine the variation produced by the selective breeding of the domestic dog with only a fraction of the enthusiasm given to Darwin's finches, one would quite reasonably assume that several new species had been produced. Even the forerunner of the domestic dog, the wolf, now accepted as a different species, will breed, and indeed has bred, with domestic dogs. The inference that evolution produces a combination of genes that will eventually prevent breeding with those of their original source is not necessarily true. So-called related species may not breed naturally because the act of reproduction is triggered off by stimuli. such as sight, smell and behaviour, often in the form of complicated courtship. Zoos frequently have problems in breeding animals in captivity when they are the same species, and these conditions are not met. When these conditions are met then reproduction is generally achieved without difficulty. Even offspring from the crossing of two different species can occur e.g. lion–tiger, swan–goose, and so on. The problem with sexual reproduction is that although it produces variation which is an advantage for survival, should the variation favour some, and therefore become a survival of the fittest, then it is these that will survive and therefore reduce the amount of naturally occurring variation. Let us assume Darwin was correct with what happened to the Galapagos finches; then if a now specialised flock of these were to be blown off course in the future, then they would be at risk. Their beaks have evolved to exploit only one food source; and since evolutionary change does not appear to work in reverse, they would have a reduced chance of survival.

Specialisation which reduces direct competition with other species for food and habitat has to be a disadvantage for the long-term survival of the species. As I have already stated, this must result in their long-term extinction, because the species are less adaptable to natural changes in their environment. There are many examples of this occurring and one has only to examine the present list of endangered species to see how these have evolved to become dependent upon the existence of certain natural and often fragile conditions for survival. Species which have not specialised are able to survive. Adaptability being the key for their survival, and not specialisation. So evolutionary change which once having occurred, and which then appears to be irreversible, may not be as advantageous as it would first appear. According to certain estimates almost all of the species that have ever lived on earth have already disappeared, and are now extinct.[1] What is interesting is that there does not appear to have been any loss of the major structural changes which have occurred during evolution. The evolution of fins, limbs, wings etc. have all been retained. It would appear that once the dinosaurs had evolved the necessary changes to produce mammals and birds, then their function had been fulfilled, and it was advantageous, in order for evolution to proceed and fulfil its aim, that they became extinct.

It may be argued that the whale, for example, having become a sea dwelling mammal lost its limbs, which had originally evolved for use on land. Many birds too have reduced wings, which had presumably evolved for flight, and are now no longer needed. Why should this be so? It does appear that organs which have evolved and are not used degenerate. It may well be that a bird may live in an environment where flight is not required for survival, but it is difficult to see how losing this ability to fly is considered advantageous.

There does appear to be a limit to the amount of genetic variability, a barrier being reached after which no further change occurs. Scientists have come across this barrier in attempting to breed beet plants to increase the sugar content, and also in the breeding of fruit flies in genetic research. This barrier must challenge the idea of constant change being possible, in order to bring about the dramatic changes needed to continue to produce new species from existing ones.

The main difficulty with deciding how life in all its forms came into existence is that the Darwinian theory is the only one we have, and as such it is generally accepted, not so much because it produces evidence but despite the evidence. It relies on the presumption that a mutation which is produced by chance will be perpetuated. The difficulty here is that in any population the mutation would have to be widespread so that in reproduction it would be well spread throughout the population, otherwise it would soon be bred out. Mutations which are the result of genes undergoing some change are now generally accepted as the basis for natural selection. As I have already discussed, should the

---

[1] Lewontin, R.C., *Doctrine of DNA*, London, Penguin, 1992.

resulting mutation be advantageous to the organism, it will increase its chance of survival. Should it prove disadvantageous then it will result in the organism having a lower survival rate and the mutation will die out. The peppered moth is frequently used to support this supposition.

The peppered moth is found in two forms, a light coloured form and a black one. Both forms were taken by collectors, and one can see from existing collections that in the early 1800s most of the population was of the dark form. This change was occurring most dramatically in the industrial north, Manchester in the UK being used to record the change. By the latter part of the twentieth century the light form was dominating the population. This change is used to demonstrate natural selection at work. The moth rests on the bark of trees during the day. In the cities the light form showed up on the soot-darkened trees of the industrial north and the dark varieties were hidden. The survival of the fittest came into play. Birds fed on the lighter-coloured ones and their numbers reduced. After the Clean Air Acts the reverse occurred and the black variety suffered more from predation.

Let us look at the evolutionary evidence. Assuming that the black mutation increased as a result of the darkened trees, then this is acceptable. However if the two forms existed in the same way as black and white humans exist, then all that occurred was a shift in the population. In the cities with darkened trees the black form of moth had an advantage, in the country it was at a disadvantage, while before industrialisation the black form would always be at a disadvantage.

Research was carried out on a large scale in the 1950s to demonstrate to the scientific communities on both sides of the Atlantic the proof of this as an example of natural selection at work. Both varieties of moth were bred in large numbers and released. Retrapping them then enabled the survival rate of the two varieties to be recorded. The work was carried out by Bernard Kettlewell. Over several seasons thousands of moths of both varieties were bred and then released in habitats differing in the coloration of the barks of the trees. The numbers of each variety of the recaptured moths were recorded. The published results were dramatic, and Kettlewell received great acclaim from the scientific community for demonstrating the way natural selection, and the survival of the fittest, could be shown working. Unfortunately all was not well. In her book *Moths and Men*, Judith Hooper shows how flawed the published results were. The numbers just did not add up. Ornithologists joined in by saying that birds had never been observed picking the moths off tree trunks. Entomologists too were critical, saying that the moths rarely settled on the trunks of trees during the day, preferring to hide in the angles of the branches. Photographs of the two forms of the moth on the trunks of the trees were shown to be dead specimens pinned to the bark. It is not the first time that so-called scientific evidence has been used to support a theory and later shown to be flawed, and I doubt if it will be the last. Despite this the peppered moth will continue to be cited as an example of natural selection at work, and the questionable photographs still used to illustrate biology textbooks.

Mutation in the genes can for example result in, say, factor A, and because factor A is advantageous it is carried forward to the next generation; but how would this factor be increased to result for example in the long neck in the giraffe? Darwinists would say by natural selection, the tallest offspring having a greater chance of survival than those with shorter necks. Let us for the present accept this as being true. However, evolution, if it is to work this way, has to do more. It has to produce factors which are not present in the genes in order for the mutation to be effective. Longer legs and other changes in both external and internal structure of the body are called for.

The arising of mutations is, I accept, inevitable. The problem is that the mutation by itself would provide no advantage as such unless it was of a trivial nature. To produce a fundamental large change, which would be required in order to result in a new type of animal, would need a whole range of changes to take place. Any new mutation arising during this changeover which did not fit into the changeover process would be a spanner in the works, and would lead to an animal less suited for survival than the original. This would appear to be the case from the fossil record which shows such changes arising quite suddenly, with little evidence of a gradual change. All the observable changes may not have been the direct result of genetic mutation. By changing one factor others may arise as a consequence. For example, increased brain size, and the resulting change in the shape of the skull, may result in changes in nose and ears. Equally there are factors which would restrict change beyond certain limits, for example the size of a bird's egg. The larger it is the thicker the shell has to be, and the more difficult it would be for the chick to break out.

I have stated that it is difficult to accept that the major changes needed to produce a new kind of animal can arise by mutation and natural selection alone. I contend that you cannot produce the major changes needed to change a member of one group into another in this way. It is the equivalent of shuffling a pack of standard playing cards and producing a new suit or a pack of tarot cards from a normal pack. This is what would have to occur to result in changing an arm or a foreleg into a wing. The Darwinists would say that the raw materials are already there, and by changing the position of the bones, manufacturing feathers instead of scales, and so on, it is quite possible for this to happen. I would be delighted to see modern technology convert my car into a boat without the addition of any new materials, with it still remaining a functional piece of transport while the entire process took place. It would also have of course the additional advantage of knowing exactly what it was trying to achieve. The Darwinists would counter this by saying evolutionary changes take millions of years. Fine, I won't place any time restrictions. They may also contend that evolution does not predetermine what the end result will be, in other words that the changes would not necessarily produce a boat. This however is the whole point of my argument. Changes in evolution resulted in the creation of the main animal groups. In other words, the production of wings resulted in birds with the ability of flight. The mutation of fins into limbs resulted in the production of amphibians with the ability to move on land. The point is that all the changes result in an animal which is an improvement on the one from which it started. Putting it simply, the fish can swim but the frog can both swim and walk; the reptile can walk but the bird can fly and walk. In view of the counter-argument, I will remove the restriction of a predetermined plan. It need not change my car into a boat. It can change it into anything it likes as long as it will end up as a vehicle that is able to travel on land and water, or on land and in the air. I will not however remove any other restriction, especially the one that ensures that it must always remain fully functional as a means of transport throughout the changeover. I will ignore the fact that it should also be competitive in its efficiency with all other cars while these changes are taking place.

Modern plant growers would be delighted to be able to grow a blue rose or a black tulip, but the gene is not present. The whole concept of genetic engineering consists of putting genes from one species into another one. This is like adding the tarot card to the pack before shuffling and redealing. Now a unique new hand can be produced, or a new organism in the lab. I cannot see how this can occur naturally by the production of a mutation and natural selection. It is hard to accept the idea that a combination of random changes will eventually lead to an advantageous effect which if it is to be of a functional value necessitates a whole range of supporting ones. It means the retention of changes that may prove useful in the future and the loss of others which may prove to be detrimental. The production of any new or modified organ would have to be achieved by following a certain sequence of events. Some biologists playing about with computer models demonstrate that this can be achieved quite easily, and new animal forms can be produced from existing ones. The criticism of this is that there is a vast difference between producing a shape on the computer screen, which the operator interprets as for example a new insect form, and the real life production of an insect from a crustacean, which would be able to carry out all of the life processes needed for its survival and reproduction, in ways which will be far more complex than the ones from which they evolved. This criticism applies to any interpretation of models used to show the way in which evolution took place. The observer intentionally or unintentionally influences the resulting changes by having a preconceived goal, for example the construction of a sentence from a jumble of letters, or the interpretation of a particular design. The problem is that the operator is conditioned to know what he is trying to produce beforehand, while in nature the organism, according to the evolutionists, would not have this privilege.

There is the old argument that if one found a watch one would reasonably assume that someone had made it, as all its components could not have come together by chance. This is of course the line of reasoning used by the creationists. It is countered with two lines of argument by the evolutionists. First that the watch had in fact evolved from simple forms of time-recording apparatus, such as sundials and sand-timers. Secondly that the building of the watch was the result of gradual advancement in knowledge and the availability of new materials. This is fine, but the point is that at every stage during the evolution of the timepiece, the designers knew what they were trying to achieve, .and each component was functional to some degree. My point is that using the evolutionists' own argument, a living organism does not apparently know that it is going to evolve an organ in order to see, or a limb which will allow it to walk. Should it be that it does know, then evolution is not just the natural selection of random variation and mutation, but a pre-set plan which determines the order of evolution. The second line of argument against creation is to accept that the watch could not have just come together by chance. The proponents of this argument use the example of a blind watchmaker. What was found could have arisen by mutation and natural selection; the point is that it just happened to finish up as a watch, in the same way as an elephant has finished up as the animal we all know. In Darwinian evolution there is no pre-set plan. Should we ever find life elsewhere in the universe, and it resembles the life forms we find on earth, then the Darwinists will have to start a rethink.

Let us suppose that I was to leave my home and just travel around with no instructions about where to go. What would be the odds of arriving for example in New York? However, if I was informed that New York was to be my ultimate destination, then even with the most difficult handicaps, no money, physical disability etc., my chances of reaching New York would be infinitely greater than if I had not the slightest idea of where I was going. The point I am making is that if evolution is to work by means of mutation and natural selection it has to know from the beginning what its objective is going to be. The die-hard Darwinists would disagree. Their argument is that one does not have to reach New York to be successful, any city will suffice, and that is the same way that blind chance has determined the course of evolution. I can accept that the creationists' theory of the origin of life is unlikely, but the evolutionists' suggestion that the increased complexity at every stage of evolution was just the result of chance is difficult to accept. Every organism is dependent on the stage before if it is to be successful. Every ecological system and food chain shows the complex relationship between the different life forms. The tree of life may not be a straight line between a unicellular organism and man, but each step is a logical sequence of events. The creationists in their six days of creation did manage to get the sequence of evolution correct.

I have so far attempted to show how the idea of evolution had been considered long before Darwin's theory, but that he was the first to sell the idea not only to fellow scientists but, more importantly, to the general public. Although he had certainly not answered all the questions, he did produce a foundation into which could be fitted new evidence, such as the role of genetics in producing variation, and more recently the role of individual genes. His theory, although in direct conflict with religious belief, did gain general acceptance, because it did much to support the social and political policies of that period. The ruling classes ruled because of their supposed superiority, and this was now substantiated by scientific evidence.

How strong was this evidence? From a theory, which despite its title, failed to provide any evidence of how species had evolved, came our present explanation of the origin of all the major classes. I have examined broadly the effectiveness of mutations in producing the changes necessary for this to occur, and the underlying suggestion that the whole process presupposes the existence of some as yet unexplained predetermined plan rather than it being the result of a series of chance occurrences. We can now look in more detail at the evidence, or lack of it, which lies at the foundation of the present evolutionary theory.

Chapter IV
THE ORIGIN OF LIFE

Although I question the evidence provided by the Darwinian theory of evolution, there is no questioning the evidence that life now exists on Planet Earth, or that there has to have been a time in the history of the world when this was not so. What is questionable is how life originated. Although much has been written on the way life has evolved from simple organisms into the more advanced life forms which culminated in the creation of the human race, and with it the generally accepted theory of evolution, little advancement has been made in explaining how life itself originated. Scientific opinions are as divided as are the various religious beliefs to account for this phenomenon. There is however this one accepted fact from which one can at least start. Life at the present time exists on Planet Earth, and the question this poses is how did this come about? Now this may have been a unique occurrence, and if this was so it cannot be repeated. It can be argued that at some time in the future our understanding of science will allow us to create life in a test tube, but this would not solve the problem. It would be the product from the creation of life that was creating it, not the original production of something which was unique. Life, if we were ever able to produce it, might help us to understand more of how it may have come about, but it would be no more original than someone studying a masterpiece in art, and then faithfully producing a replica. By attempting to mimic the conditions which existed on earth during the period when life seems to have originated, and by supplying the ingredients which would have been needed for the manufacture of living matter, then certain amino acids have been produced in the laboratory. This work, by Stanley Miller in the 1950s, led to great enthusiasm that life could be created in a test tube. However, amino acids are not alive, and no nucleotides have ever been produced experimentally, and so without these there can of course be no replication. The conditions too which supposedly existed on earth are to say the least speculative, and so for that matter are the presence of the ingredients needed to produce the resulting amino acids.

Life arose and may still be spontaneously doing so. This is something which science, albeit reluctantly, has had to accept. Louis Pasteur had had great difficulty in convincing scientists that life can only come from life, when explaining decay and disease by the action of bacteria and other micro-organisms. Decay is easily observable in everyday life. Organic material such as food is broken down and becomes inedible. Living material once dead quickly decays and is recycled. Micro-organisms were shown to be present, and it was assumed that these resulted from the breakdown rather than being the cause. Pasteur was convinced this was not so. By boiling meat broth in a flask, and then sealing it to prevent the entry of air, he showed that it remained fresh. When the seal was broken and air was allowed to enter, the broth quickly broke down and micro-organisms were shown to be present. By repeating this in the clean air of the Alps and then in the streets of Paris, he showed that it was not the air but the dust and dirt in the air which transported the micro-organisms. His experiments proved that life only comes from life and does not arise spontaneously. Now science has to conclude in its acceptance of evolution that life did arise spontaneously from non-living material.

The origin of life follows the principle that there was a gradual change from simple forms to more complex ones. The difficulty with this is where to start. Even the simplest life forms, such as bacteria, are complex. They follow the same biochemical plan of all living things, yet we are led to believe that from the coming together of a self-replicating molecule, the simple cell evolved. There is no evidence that this is true. It just seems to be the only explanation on offer. Of course if it can be shown that life has evolved on other planets, then life and its origin would not be unique. Until such time, there is the possibility that it has only occurred the once, and this limits scientific investigation. It has been suggested by one Nobel science prize-winner that life did not in fact originate on earth but somewhere else in the universe, and that the earth was seeded either intentionally or by chance with primitive life in the dust from passing comets. There is no evidence of this happening, but it is interesting that it is offered as a more likely explanation than for life to have evolved by chance on the lifeless Planet Earth. This is however really only passing the buck, since at some time it has to have arisen somewhere, if not on earth then on some other planet. Accidental contamination of other planets with simple life forms from earth are a serious consideration in our own exploration of other planets. Although still in the category of science fiction it is still a possibility that this may have occurred in the past, and we are the indirect result of such an intended or unintended contamination.

In order to make some progress we must attempt to answer the first question, that is, whether life does exist on other planets in the universe or are we in fact alone? Next we must ask if the creation of life was an inevitable consequence of the birth of the universe, or was it a complete fluke which may well have happened just the once? Should it prove to be the former then it would appear that we need more than our present understanding of the natural laws to explain how this could come about. Using our own knowledge and experience of utilising these laws to reach our present state of scientific and technological ability, we have been unable to even come close to achieving the synthesis of even the basic building blocks of life, let alone producing a simple self-replicating organism. Physical laws provide predictability and rigidity in the way they operate. We can for example predict how long it will be before our own sun, and with it the earth, ceases to exist. We cannot however predict with any certainty how the human race will have changed within the next few generations. Living matter defies predictability, and therefore it would appear that there are other factors at work which can explain this fact. A chemist can predict what will occur when two chemicals are combined together; the result will always be the same. The behaviour of complex living organisms does not afford the scientist this luxury.

It may be said that we are now in the dark area of asking unanswerable questions, which in our attempt to fathom what may in fact prove to be unfathomable should be ignored, since it is unlikely that an answer can ever be found. It is the classical question of 'If God made everything then who made God?' Science has to work within the man made boundaries of logic, time and probability. We have to decide what is possible and what is not possible in a universe where such limits have to be meaningless. Although the events needed for life to begin may seem so impossible, if we accept that the universe is infinite in both space and time, then anything must indeed be possible. So the first organism capable of reproduction may have occurred by chance, but this can never be proved. Should life however be common in the universe then one might well suspect that some other factors were involved apart from chance.

For life to have arisen spontaneously as a result of chance, and with only our present knowledge of the universal laws to assist us, then we must consider the odds. For just a single protein molecule to be produced this way, the odds are reckoned to be $10^{40,000}$ to 1. Even accepting the vastness of the universe and the number of planets that might act as the assembly bench for such an occurrence, the odds are still incredibly high. To produce the genetic code which is essential for the primitive life forms to be able to reproduce and evolve appears to be impossible. The complexity of living organisms, however primitive they may be, cannot be over-exaggerated. Not only have we to account for how the materials needed arose, but how the organisation and interrelated functioning of the components arose. It is the interaction of these components which produces life. DNA is not alive, nor are the molecules which produce it. Only the exact combination of the millions of molecules which make up a living cell produces life. More than that, each molecule must play its own individual part to produce a living organism. Life requires hundreds of thousands of proteins, each made up of amino acids, each in its own position, and each manufactured from the formula contained in the DNA.

The formation of the DNA, the genetic code, creates a problem. It can only function at one level, and it is difficult to see how it could have evolved from a simpler form. Any change, however small, would affect the production of proteins, and this would make the function of the DNA impossible. Most biologists and indeed most other scientists reject the idea that there appears to be purpose. Yet purpose does exist in living organisms; even if not in all, then certainly in human behaviour. Since we are all the products of what occurs in our cells, then when did purpose arise, and from what? When we then consider the sequence of events which is necessary to produce a human being with the mind and ability to produce the technology of today, then it seems unlikely that this could occur anywhere else in the universe. Certain scientists have pointed out that if the whole tape of recorded evolution was to be wiped clean, and allowed to start again, then it would be impossible to produce the world we have today. Any person reading these words is only doing so as a result of countless chance occurrences in their ancestral history, which have resulted in them being here at this point in time. Despite this, and since our solar system is so much younger than the rest of the universe, there is a belief that if life had evolved on other planets it would by now have evolved into advanced technological civilisations. As a result of this thinking, great effort is being made to find, and also to then contact via radio waves, any such civilisations. The question I find more interesting is not whether there is life out there or whether we are indeed alone in the universe, but why we have evolved to want to know the answers to these questions?

It would appear that the appearance of life on this planet occurred immediately after the primitive earth had cooled. What is of interest is how many times this phenomenon may have occurred. With few exceptions all life carries the same arbitrary coding. One might have expected different codes to have arisen, in much the same way as different languages arose in the human population. Evolutionists would of course argue that different life forms may have arisen, but natural selection ensured that only the most adept would survive. Again there is no proof, only speculation. The world is a big place and if self-replicating life forms arose in different parts on the earth, then it seems strange that they have all followed the same pattern of development, since virtually all life forms are based on the same DNA genetic structure. We can look at it another way, that is that the future of life on this planet is not permanent. Even if one ignores the future effects of new diseases, self-destruction, loss of natural resources, meteorite or comet collision etc., one is still left with the fact that life on this planet is limited to the time until it is destroyed by the death of our own sun. Agreed that this is estimated at about another 900 million years from now, but if all but the more primitive forms of life were to be destroyed before then, would there be sufficient time for evolution to start again from scratch? It might be argued that because of the uncertain future of the earth, life and its evolution may well have achieved nothing. On the other hand it could be that life will have evolved to gradually occupy the whole universe, and evolved the necessary ability to lift the veil of ignorance and attain total enlightenment. Should that be the case it may well be that we are just part of the spread of life throughout the universe and not necessarily the origin of it. A farmer sowing seeds would not expect them all to germinate, and of those that did, not all would flourish long enough to be harvested.

The knowledge that an individual will not be around to know the truth can act as a brake on one's endeavour to seek an understanding of man's place in the universe. Perhaps the safeguard to this is that our thoughts and behaviour are the results of our genes, and belief in God arose in order to allow us to come to terms with the inevitability of our own deaths. Science and religion have to be linked, since neither at the moment can provide the answers to our questions. Religions are a universal factor in various forms, and their origin and acceptance must be as much the product of our genes as is any other type of behaviour. Science offers no explanation for our purpose or existence. Religion offers two alternatives, with no real evidence for their validity. Either God set up the universe in the beginning, and then had nothing more to do with it, or God sustains the universe and everything in it.

Our present scientific explanation for life is difficult to accept. It is rather like the national lottery, in that it is possible to pick out six numbers and win, but to keep on doing so would suggest that more than just chance was involved. The difficulty lies in the fact that we can construct a complex machine by first designing it, then drawing up a list of the components needed, obtaining them, and then building it. Evolution however cannot work that way if we assume that there is no plan or view of the finished article. The other problem is that machines are needed to make the components. In nature proteins are needed to make catalysts, and these are needed to make proteins. How can step by step evolution occur when each part is dependent upon another part for its existence? In highly developed multicellular organisms the same applies. The digestive system depends upon the transport system, which in turn depends upon the nervous system, which depends upon an excretory system. Each system has to evolve together, with the others in order for the organism to function as a whole. I can build a machine, and when I have finished switch it on and it may work, but the evolving organism has to be able to live and function as it is being built. Each of the systems functioning in a living organism is needed in order for it to live. Controlled energy, for example, is required in order to create a system capable of producing energy.

Returning to the question of the origin of life on earth, we cannot be certain that life did in fact originate here. Let us suppose that simple life forms had evolved on Mars or some other planet, then they may have arrived on earth via meteorites, and since conditions here were more suitable for evolution they evolved into all the different life forms that we have on earth. Equally it may be said that if simple life forms are found to exist on Mars, or had existed on Mars in the past, may not their origin have resulted from similar contamination from earth? Should this be the case then it certainly reduces the chance of finding advanced life forms elsewhere in the universe. The apparent lack of life elsewhere in our own solar system suggests that conditions determine the development of life forms, rather than life adapting to the conditions.

Apart from the possible part played by meteorites, comets too may well play a part in the spread of life forms throughout the universe. We do know that simple life forms do have the ability to withstand the conditions of outer space, so comets may well have had some part to play. Life forms have now been found thriving in the most extreme environments on earth; from the high temperature of vents in the deep ocean floor, to miles beneath the earth's surface, as well as below the ice caps. What is important is that these organisms are able to utilise chemical energy and are thus not dependent upon the sun's energy. It may be that these organisms simply colonised these habitats from their origins on the earth's surface. Equally it may be that they originated there and spread to the surface of the earth as the conditions for evolution and subsequent expansion stabilised. It does appear that life appeared on the earth at a very early time in the earth's history.

It is from all of these possibilities that opinions and theories are born. It can well be argued that they all have equal merit, because the simple fact is that we just don't know. Evolution has brought us a long way in our understanding of the universe, and since the function of evolution is to increase the ability of survival, then this understanding must be achieved if we are to survive in the future. I have stated that it is difficult to see how life could have evolved into its present state by means of our present understanding of physical laws. Quantum physics working at the sub-atomic level may well have a part to play in explaining the origin of life, and even assist in our understanding of its purpose for its subsequent evolution. All life forms up to man have a specific function in supporting him in his quest for seeking the purpose of the universe as a whole. As such I cannot accept that all of this results from the chance coming together of certain specific substances.

I have already discussed the many difficulties we face in attempting to explain the origin of a functional cell. It creates many nagging problems, just one of which is that a permeable membrane has to evolve in order to contain the functional parts of the cell. It is not sufficient to state that the chance coming together of simple substances needed to produce complex chemical compounds can result in the production of the materials needed to produce the chemicals of life. These substances must have been constructed so as to allow a simple life form to be produced that is fully functional. To deny this would be the equivalent of saying that the substances needed to produce a single brick could arise by chance; but then as a result a fully operational chemical factory could suddenly come into being, with the added factor that it would have the ability to reproduce further factories. The problem is that none of the complex processes going on in the cell can function in isolation, but only in conjunction with other systems.

To return to the cell membrane. Without it the cell contents cannot be contained, and unless these contents are fully functional, the cell membrane cannot arise. It can be argued that the cell membrane arose as a result of certain chemicals coming together by chance; after all, membraneous material can be produced quite easily in a test tube. The point is, however, that when a cell reproduces, then it also reproduces a cell membrane. In other words, the instructions for its manufacture must already have been coded in the DNA.

The complexity of the human body frequently creates a certain awe amongst even the most non-scientific of people, and rightly so. What is generally overlooked is the complexity of a single cell. Here organs are replaced by organelles, which are concerned with carrying out most of the processes associated with a multicellular animal such as ourselves. I do not intend discussing these in any great detail, since to attempt to do so would take a whole book. Certain facts should however be taken into account when considering how single cells came into existence. The cells of all living organisms are made up of about twenty-five different amino acids, nitrogenous bases, lipids and sugars. The most important is the nucleic acid, or RNA and DNA, which are responsible for the activity and reproduction of the cell. Energy essential for life is obtained and controlled by the use of adenosine triphosphate, or ATP, and this substance is identified in all living organisms. The cell membrane is a complex structure, and is different in each type of cell. Its function is not only having to contain the contents of the cell but also to selectively allow materials to enter and leave. The nucleus of the cell which contains the genetic instructions has its own separate membrane. Since all forms of life, independent of the type of organism, start life from a single cell, it therefore has not only to be able to reproduce itself, but also to be capable of differentiating into any of the different types of cells found in multicellular organisms. These have different structures and functions. How this is achieved is not fully understood.

The genetic code is the key to the system upon which everything else depends. It decides the structure, actions, reactions and function of every cell. How did it originate? It cannot have evolved gradually. It must have existed from the beginning as a fully functional piece of machinery capable of satisfying the requirements of the cell, and also having the ability to reproduce and evolve in complexity.

Let us consider what we are discussing. We are concerned with the control centre of the cell, the nucleus. Long molecules of DNA and protein make up the chromosomes. The DNA is made up of structures called nucleotides, which are linked together in chains. Each nucleotide is composed of a deoxyribose sugar, a phosphate group, and a nitrogen-containing base. There are four bases, adenine, guanine, thymine and cytosine. Two such strands of DNA are held together by hydrogen bonds between the pairs. The pairing of the bases is specific. The whole structure is coiled and forms a double helix. It is a stable molecule and is capable of self-replication. The genetic meaning depends upon the sequence and positioning of the bases. As an example, using the first letters of the four bases, then TCA is the genetic code for the amino acid serine. Instructions and sequences run into millions, running from one chromosome to another. The number of chromosomes varies in different species; man has twenty-three, a fruit fly four, and a goldfish forty-seven. Only about ten percent of the sequences of the chromosomes appear to cause anything to happen. The function of the other ninety per cent is unknown. In replication copying errors do occur, and these result in a mutation. However only if the sex cells have this mutation can it be passed on to the next generation. This last statement is significant. What it means is that in the millions of cells making up a complex living organism, a mistake in the copying process is only going to have any evolutionary effect if it takes place in the production of a sex cell, and even then it would have to provide some beneficial effect.

Not only is the origin for the replication of the DNA a puzzle but so also is the production of proteins. RNA has the task of selecting and assembling the amino acids which are needed to produce enzymes. These are required in the reactions to produce the specific proteins needed by the organism. The complete amino acid sequence of several enzymes has been determined, and by the use of X-ray crystallography, we have discovered their exact three-dimensional molecular structure, but no one has been able to synthesise a protein molecule in the lab.

These complex molecules do not just assemble themselves by chance. Something else is required. As previously compared, it is the equivalent of determining the structure of a brick but being unable to make one, and then trying to determine how a house is built. The difficulty is that even accepting the development of a single cell into an elephant or an oak tree, there has to be a process of control. Where does this come from? Each cell in its development and growth is not acting in isolation but is working together with others. How can this cooperation have come about?

The other significant fact is that a multicellular organism starts off as one cell which reproduces into two, then four, eight, sixteen and so on. Certain cells then differentiate to become muscle cells, skin cells and any of the others that make up the different tissues which combine to produce the organs and systems required by every living multicellular organism. So each cell is carrying the blueprint for the whole organism. It is also programmed to produce not only any of the specific cells but also to know when the required number has been produced, as well as maintaining the required amount not only to reach a predetermined growth but also to replace those which die or are destroyed.

It is known that DNA emits photons which correspond to visual light, and as long ago as 1923 it was suggested by the scientist Alexader Giervich that this could be how cells are able to communicate. This may or may not be true, but it does show the void which is created in the failure to explain how the control of the whole organism is achieved by the function of the DNA. The so-called junk DNA which forms a large part of the DNA coding is written off as being functionless. This is surprising when one considers the critical influence and control attributed to the rest of the genetic coding. Could this so-called junk DNA be the raw material available to be programmed so as to take on new roles?

The structure of the brain in humans and our ability to produce consciousness has to be credited to the ability of the DNA coding to produce certain proteins which would allow this to happen. This must show that the genes are capable of responding to the external needs of the organism. We are informed that genes are switched on and off to produce these results. Here one is reminded of Pavlov and his dogs. By feeding his dogs in conjunction with the ringing of a bell they became conditioned to associate the bell with food and would salivate at the sound of the bell even when no food was present. The secretory cells producing saliva had to be able to respond to this change. Before this conditioning occurred, the changes from the response of the sensory nerves of the eyes to the visual sight of food had to be replaced by those responding to the input of sound from the ringing of the bell. The nervous system's response was to send the information to the salivary glands to instigate the production and release of saliva. Such a significant change of behaviour would from an evolutionist view be the result of the production of a mutation, and subsequent natural selection. The response was however the product of learning, and the consequent production of a new response by the cells and organs concerned only occurred after the conditioning had taken place. The long-winded way of producing change by natural selection would take too long to spread through the population for it to be effective. The question posed is, can the new learned behaviour, resulting from a change in the stimulus needed to trigger reflex behaviour, be retained and passed on? This is of course an echo of Lamarckism, which will be examined later. Critics would say that the genes are not involved here, the brain simply receives a stimulus from the sense organs and this triggers a response to the stimulus via the nervous system. All behaviour cannot be explained in this way. The brain is for example already conditioned before birth to respond to certain stimuli such as pheromones, which will produce a certain specific response. Mammals when they are born suckle. This behaviour is not learnt, therefore it has to have evolved with the formation of the brain. What therefore is its origin? One has to remember that this information has to have been present in the single cell produced at the time of conception.

It is hard to see how it could be the result of many years of chance mutation and natural selection, which had also to be synchronised with milk production by the mother. Simple reflex behaviour is a common factor in most living organisms, and one may be tempted to accept that it had evolved in the same way that other beneficial factors have been taken aboard. Complicated behaviour however such as nest-building in birds is another issue. Here it has to have been learnt and modified by experience and this experience passed on to its offspring. As we have already seen, conditioned reflex shows that inborn simple reflexes can be changed quickly.

Let us look at this phenomenon a little closer. Simple reflex behaviour is not learnt but is passed on via the genes from one generation to the next. It has an obvious survival advantage and therefore it can be argued has gradually evolved and passed on using the basis of survival of the fittest. Should we however examine the conditioned reflex then we face a problem of how this can be similarly explained. Let us consider the implications of some examples. Should a horse be placed for the first time in a field near a railway line, it will be startled when a train passes by. This is a straightforward reflex response to the noise. However, after this has occurred several times the horse will cease to respond. The horse is of course an intelligent animal and may have simply decided not to respond. We can test if this is true by observing a garden snail at rest with its horns extended, a vibration on the ground close by will cause it to withdraw its horns. After this has been repeated several times it will ignore the vibration and its horns will remain extended. It has been conditioned to the stimulus. How has this simple reflex behaviour changed? Acquired behaviour, we are told, cannot be passed on to the genes. Its origin lies through chance mutation being passed on over a period of time, with each gradual change providing some advantage to the animal. How then can one explain how it can be changed so quickly? With highly developed animals the behaviour could be explained as conscious behaviour, but even simple freshwater microscopic animals are also able to be conditioned to non-reactive instinctive behaviour.

As I have previously stated, DNA is not alive. As I write these words I am looking at a scar on my hand. A few weeks ago a chisel I was using slipped and it gouged out a piece of flesh. Somehow the genetic coding knew what had happened. Instructions sent out ensured that the ever-present defence system of my body came into operation. The bleeding stopped, damaged cells were removed and any bacteria destroyed. All this the result of increased efficiency by generations of natural selection! The replication of the undamaged cells now went into overdrive to replace the missing tissue and the injury healed. There is nothing unusual in this story. However, if some of the evolutionists who explain so easily the way evolution functions and how life evolved can explain how the information of the injury could be transferred to the DNA in the undamaged cells, and for them to also know when the damage had been repaired, then I may be inclined to accept their beliefs with more confidence.

DNA controls the activity of the cell, but that is all. This fact tends to be overlooked in the general enthusiasm for its function in living organisms. In order to function DNA must be part of a large team of molecules, which begs the question of how it could have arisen in the first place. What we have is non-living materials which are controlled by natural laws and behave in accordance with these laws, and living organisms which as they increase in complexity have the ability to lose the predictability which laws produce. The problem is trying to bridge the gap between non-living material and living material.

Let me pose another question, which is that if life came into existence by chance, then why is its functional purpose survival and evolution? Could it be that its purpose is to produce a conscious being the function of which is to produce more evolved and conscious beings? Since evolution has led to the development of the human mind so far in advance of our so-called nearest relatives, the apes, then it would seem that consciousness only arises when a certain physical complexity has been reached. The whole system of life is above all else based on the driving force of reproduction.

Simple one-cell animals usually reproduce asexually by division, forming two daughter cells which are generally identical to the mother cell. This produces very little variation of the offspring, and as there is little or no variation there is a reduction in the organism's ability to adapt to changes in its environment. Some one-cell animals do however at times fuse together, resulting in their genetic material combining before division takes place. Here is the origin of sexual reproduction, with its increased likelihood of the production of variation. Multicellular animals reproduce sexually by the production of specialised sex cells or gametes which are designed to be unable to fuse with themselves but only with those from another individual. This is achieved by the production of male and female organisms. The necessity to produce variation dominates all branches of life, both in the animal and the plant world. The only things which differ are the methods employed: external fertilisation, internal fertilisation, and in the case of flowering plants the use of vectors such as insects, in order for this to be accomplished.

In animals such as fish the male and female gametes can be produced and then released into the water in large quantities for fertilisation to take place. The problem for reproduction in land-dwelling animals had to be overcome and this was to lead to the ultimate internal fertilisation and development of the embryo in mammals. We have to consider that all of these changes from egg-laying to mammalian reproduction were the result of chance mutations of the genes at each stage to provide the functional changes required. More importantly was the production of hormones required to instigate the behaviour of the animals to carry out the process leading to the production of new offspring. Apart from behaviour, inter-recognition of the sexes necessitated changes in size, body shape and colour as well as adaptation of sounds such as specific mating calls. Without these changes reproduction would not occur since it works against the principle of the survival of the individual. Let us accept that the chance changes resulting from genetic mutations may provide the organism with certain survival advantages. How can sexual reproduction have arisen when it provides no immediate advantage to the individual? It has led to the situation where the whole purpose of a living organism is for it to mature, pass on its genes and then die. How does this fit into the concept of genetic changes increasing the survival chances of the individual? What I am saying is that in order to produce offspring the survival of the individual is reduced. This does not suggest to me that the process of evolution is the result of chance changes in the genes which provided some survival advantage to the individual, but a predetermined programme of events which would ensure that the whole sequence of evolutionary development would occur. The sex drive in the higher animals has become the most important factor for their existence, arising as it supposedly did from the chance combination of material that had the ability to replicate. From this arose the concepts of purpose, direction, control and function. These are the products of design. Yet we are expected to believe that evolution can be explained by a continuing series of events that once started, follow blindly a set of natural laws that govern the whole universe.

Education and ambition determine that we are in a better position to marry and raise children. The driving force is to place us in a position where our offspring will have the best chance to mature and be in the position to raise the next generation. Once this is achieved we deteriorate and die. One might argue that the role of the genes is to serve the best interests of the organism, but here we have a system geared to serve the genes themselves. The living organism, be it bacteria or man, is merely the vehicle for the reproduction of the genes.

All living organisms including man are programmed to reach sexual maturity, reproduce and then die, having passed on their genes to continue the process. Why should this be so if the process arose by chance and is without purpose? This of course raises the bigger issue, that of the origin of the universe itself. Using a theological slant, it would be logical that conscious life had to evolve in order to justify the existence of the universe, which without the existence of life to be aware of its existence would be meaningless. I am not suggesting that this is necessarily correct, but it is a point of view which due to our present ignorance cannot be ignored. It would appear that evolution is on course to remove this ignorance, and therefore it would be wise to leave all options open. Evolution has after all led me to believe this may be true, and a great deal of natural selection must have taken place for me to have reached this conclusion.

The truth is that at the moment we don't know the answers to the questions we ask. We have a choice of answers to choose from, but have no way of knowing which of these is correct. The danger is that once we believe that we have found the answers then we stop asking the questions. Darwin formed his theory without the knowledge of genetics, and later their importance added weight to his theory. Perhaps the religious concept of a creative God responsible for the universe and life will receive similar support in the future. I certainly feel that the evolution of life needs some explanation, and see no reason why the two main opposing bodies of opinion would be unable to unite in an attempt to seek the truth. Either evolution has no overall design or plan, or else the intended goal of evolution was, and still is, to produce minds capable of determining the truth, as well as realising our ultimate destiny.

Let me now recap on the points I have covered. I have looked at the improbability of life evolving from non-living material and the complexity of even the simplest of life forms. There is of course no actual evidence that life did in fact originate on earth, and it could well have been the result of contamination from some other source. I advise anyone to consult any text book which deals with the detailed structure and function of the genetic code and the role of DNA and RNA, and consider that this supposedly resulted from the chance reactions of certain molecules that were then able to replicate themselves. Genes are able to programme protein synthesis via genetic messages in the form of RNA. The genes' unique sequence of DNA, the nucleotides, provide a template for the assembly of an equally unique sequence of an RNA nucleotide. From this information the amino acids are assembled, again in the correct sequence, to produce the required protein. For this to occur enzymes are in fact required, and here lies the problem. Enzymes are themselves proteins. So we need enzymes to manufacture proteins, and proteins to manufacture enzymes.

Lewis Thomas in his book *Late Night Thoughts* writes, 'DNA is no longer a straightforward set of instructions on tape. There are long strips of what seems nonsense between the genes edited out for the assembly of proteins but essential nevertheless for the process of assembly. Some genes are called jumping genes, moving from one segment of DNA to another, rearranging the message, achieving instantly a degree of variability that we once thought would require aeons of evolution. The cell membrane is no longer a simple skin for the cell, it is a fluid mosaic, a sea of essential mobile signals, an organ in itself. Cells communicate with one another, exchange messages like bees in a hive, regulate one another. Genes are switched on, switched off by molecules from outside whose nature is a mystery. Somewhere inside are switches which when thrown one way or the other can transform any normal cell into a cancer cell, and sometimes back again.'

Even the simplest of organisms have this complexity, and even if one accepts that by chance certain molecules have the ability to combine and then replicate, this does not explain how such complex systems followed. Much is made of the fact that the driving force of such complex development was the need to compete and be able to survive in new environments. The expansion of life from the sea on to the land and into the air could not be just the result of the quest for food. The move to different environments would present too great a demand on the changes needed to compensate for any benefits obtained. I suggest that this basic drive cannot be explained so easily. It is still with us today in our quest for space exploration and our talk of the colonisation of other planets in the future.

I have also considered the fact that behaviour in response to the environment has to occur in living organisms in order to survive. We are informed that genes are switched on and off as a result of these signals from the environment, and this results in a behavioural response to the stimulus. I contend that simple instinctive behaviour may well have evolved in this way, but what of complicated behaviour which results from acquired experience? How was this incorporated into the fixed pattern of the genes? Reproduction is the most important factor in all living organisms, and everything is geared to ensure that it takes place. It would appear that the sole function of all living organisms, including man, is to mature, produce offspring and then die. This process is of course essential for ensuring that evolution occurs, but this again points to a purposeful pattern, not one which has arisen by chance. From the concept of the survival of the fittest, reproduction places any individual organism at a disadvantage. What therefore, I ask, is the survival advantage of the sex drive? Could it be that the genes are programmed to produce offspring that can benefit from the accumulated knowledge of previous generations and thus lead to a controlled destiny for the human race? Could it in fact be that living organisms are merely the machine which the genes produce and utilise in order for them to change and evolve into some preconceived goal? Before these two suppositions are rejected as nonsense it is worth considering that the most sophisticated computers have come into existence as a direct consequence of genetic evolution. In the human race the finishing line is not even in sight.

Chapter V
NATURAL SELECTION, VARIATION AND MUTATION

Variation in living organisms is important for survival. It results from sexual reproduction, whereby the organism is designed from the genetic material from the two parents. The combination of these two sets of genes ensures that the offspring will have varied characteristics inherited not only from its parents, but also from their ancestors. Much is made of the part played by variation, and by its effect upon evolution. There can be little argument over the fact that variation must improve the survival chances of a species. The fact that they are all different means that some will have a better chance than others if the conditions for survival change. Should they all be the same, then they would either all survive, or all perish. The variation also allows for certain members to be able to exploit certain resources. Human beings exhibit variation which allows for the functioning of a structured society with its division of labour etc. We don't all eat the same foods or are all equipped with the same aptitudes and skills. Were we all the same, it is difficult to see how civilisation could have come about. Although variation is obvious in some species it is less obvious in others. Wild rabbits may look alike, but since they are the product of sexual reproduction then they cannot all be the same. The variation in rabbits, which one would expect to increase their survival chances, did just that by decreasing the effect of the myxomatosis virus introduced to control the rabbit population. Most rabbits were susceptible to the virus and died, but the small percentage that because of variation survived, were immune to the effect, and carried on reproducing. The rabbit population was soon back to normal. Had the rabbits all have been the same, then they would either all have survived or all would have been wiped out. Human variation has enabled us to survive similar effects of potentially fatal diseases, such as outbreaks of influenza, and the black death. Even more recently it has been shown that certain individuals have a natural immunity to the AIDS virus. Although the variations produced by sexual reproduction may be essential for survival, I consider the resulting changes produced are still too trivial to produce the changes required for major evolution to take place. One can therefore accept that variation will result in a population exhibiting differences in skin or fur colour, resistance to certain diseases, and for example differences in behaviour. I can also agree that variation occurring in a group of organisms will favour some over others. What is important is that these variations act as a safety mechanism against changes which are outside the organism's control. The safety behaviour of hedgehogs to roll up as a protection from approaching danger works well against predators in the wild, whereas in the crossing of busy roads it proves disastrous. Those members of the species who have less of a tendency to roll up stand a better chance of surviving, and of passing this trait on to the next generation.

However, without mutations of the genes in the gametes, which in turn would alter the composition of the genes being passed on by reproduction, no new and lasting variation can occur. Without mutation, dominant and recessive genes will simply combine and produce characteristics which are, or at some time, have been present in the population. Different genes which are neither dominant nor recessive will combine to produce characteristics not evident in the parents. For example a red rose and a white rose crossed when neither the red nor the white gene is dominant will produce all pink roses. These are called F1 hybrids by growers. However if the two pink roses are crossed the F2 generation will produce red and white roses, as well as pink ones.

So if normal variation produced by normal genes combining cannot account for lasting and effective changes, then one has to accept that it is only the mutations of genes in the gametes that will result in unique changes in the genetic make-up of the organism. That may be acceptable, but what I find unacceptable is that complete changes in for example reproductive methods, such as from egg-laying by reptiles to the complex mammalian method of reproduction, can be achieved by a series of steps. The change can only function as a whole or not at all.

What we are left with is that there was a sudden appearance of adaptive structures, and the gradual change which Darwinism demands. However, if these were the result of a massive macro-mutation of genes, then it would seem unlikely that this would produce all the beneficial changes which have arisen; the creation for example of a lung, of limbs, or of feathers. Is it not more likely to have produced effects which would be detrimental and would be disastrous to the species? I have yet to hear of any evidence or suggestion that the mutation of genes on a grand scale i.e. from nuclear fallout, or excess radiation from the sun, provides any expectation of it resulting in changes which would provide an advantage to the individuals affected over those who were not affected. It is suggested that if the mutant was dominant, and proved to be disadvantageous, then it would die out by the action of natural selection. Should it however be recessive it could remain within the population until such time as it proved advantageous for survival, and then it would manifest itself and flourish in the population. We are now extending the odds of this happening. First we have to have a mutation occurring in the gametes so that it can be passed on. Secondly it has preferably to be recessive, and thirdly it can only prove advantageous if external factors occur which make it so. This pattern has to be repeated over and over again until the final change is complete. We are, I submit, back again to winning the lottery every week.

The difficulty lies with the production of species. Should species arise by the gradual accumulation of beneficial changes, and these changes are not predesigned, then it is hard to see how they can have resulted in the production of new organs and new organisms. On the other hand, if they are predesigned, in other words evolution knew from the beginning where it was going, then this raises the question of how this could be so? Some Darwinists are quick to point out that by random changes in the composition of a group of random letters, a sentence can be produced after only a few lines. The snag with this argument is that the operator knows what is a beneficial change and one worth retaining in the next line, and the one to reject. The change in the genetic code does not have this advantage. It could in fact retain a change which had no future benefit, and might easily reject one that had. There is also the possibility that if mutations arise spontaneously, then they might well mutate a gene which would have been worth retaining.

It should be pointed out that the retention or loss of a gene depends upon its effect on the organism's ability to compete. Unless the change in the gene produces some adverse effect upon the organism's ability to compete, then it will be retained. The problem here is that the organism's very existence shows that it can compete, and any change has to create an improvement in its efficiency to survive, or by the consequence of natural selection it will die out. In order to produce radical changes from one organ into another by a series of steps, each step must result in an increase in efficiency of the organ or it will be lost. To change a fin into a leg, or a leg into a wing, each genetic change has to ensure that there is no loss in its functional efficiency through each stage, or by natural selection the genetic change will be lost.

Much has been made of the development of the eye. It certainly worried Darwin, but present-day Darwinists either ignore it or explain its development by a gradual series of beneficial changes. It is certainly the result of many changes before it is fully functional, each of these changes having to be retained in the hope that further mutations will occur which will provide the next step. The Darwinists would argue that any development would provide some benefit. This I may accept, but why should mutations follow the correct order so as to produce a more functional organ? There would be no point in producing an eyelid before the eye, or an iris before a lens. Each step has to be in the correct order; that an area of skin sensitive to light may provide an animal with some advantage is I agree acceptable, but why should this stimulate a continuation of the production of mutations in order to produce an efficient eye? We must not forget that these changes have to occur in the gametes in order for them to be passed on. The production of an indent to start the evolution of an eye socket occurs with the development of second one in just the right position on the head. This is followed by the development of a pair of membranes which will in time evolve into lenses complete with pinhole apertures, and later the iris. This does not happen anywhere else on the body, although the development of an eye at the back of the head would be as much of an advantage as a rear view mirror in a car. No, they are always at the front, the result of a mutant gene that was probably responsible for the production of cells which were sensitive to light. Well, if this was how it started then it certainly got carried away. Mistakes in replication assisted in the production of retinae, rods, cones, the optic nerves with the development of the brain in order to make sense of the impulses, and a nervous system to respond. The skull also had to change in order for it to accommodate the eyes, and produce eye sockets with the development of eyelashes, eyelids, and tear ducts. I will not dwell upon the complexity of the chemical and physiological events which had to have taken place in order to convert the image into electrical impulses which could be transported to the brain, and then converted into an image in the mind. This image, it should be remembered, is also stored for future reference. It should also be noted that we don't have to actually use our eyes to conjure up an image. Think of a banana, and you produce an image of one in the mind. The function of our eyes is to transfer what exists into just such an image. I suggest that anyone should consult a biology textbook, and having read how this is achieved consider again the proposal that it is all the result of chance step-by-step changes. All this happening twice to produce two eyes. Predators of course have eyes at the front, while herbivores have them at the side. Did they gradually migrate after they had evolved, or was this a factor which came into effect as they evolved? It must be remembered that any change in one part of the body has to coincide with changes in other parts of the body, in order to accommodate that change. This means that the genes responsible for these supporting changes have also had to be changed. What stimulates the genetic instructions in these genes to be changed in order that the mutant's effect can be functionally beneficial?

Richard Dawkins in his discussion of the evolution of the eye states that it has evolved between forty to sixty times in many different invertebrate groups. Surely this would support the concept of a predetermined plan, rather than the view he favours that they were the results of the chance production of genetic mutation. Once again the setting up of a computer model to show how this could occur by chance mutation is misleading and wrong. The operator selects each slight change needed in the process leading up to the production of the eye, and rejects any others that fails to do so. Why should this occur in nature unless it had been programmed to attain the production of the eyes? The fact that it has occurred so many times in the invertebrates would suggest that it is a programme that is designed to happen. Scientists agree that between the production of mutations there is no bias. Although Dawkins rejects the idea of a programmed DNA code, he himself is playing the role of a creator. The computer operator has a definite goal but rejects any such goal in the evolutionary process. Dawkins surely demonstrates that intelligence is required to produce a preconceived result, not chance.

Each generation starts from the fertilisation of an egg by a sperm. The resulting single cell, the zygote, then develops into the whole organism. So even if evolution did happen in the way Darwinists claim then we still face the difficulty of explaining how cells are able to form patterns of tissues and integrate with one another to produce complex organs. The cells cannot work independently of each other as they do in unicellular organisms. They have to be controlled either from a separate source or by some form of intercellular communication. Cells on the outside of tissue generally have to be able to reproduce at a faster rate than those on the inside, in order to replace those lost by wear and tear. It is also apparent that from a single mother cell, specialised cells have to develop, for example muscle cells, nerve cells, secretory cells etc. This capability has to be present in every cell, but how does an individual cell know what its role is in the production of the organism? Equally, what triggers off the response of the genes concerned with this process? It is worth considering that for a cell to become a muscle cell, or any other specialised cell, it would from an evolutionary standpoint have to be explained by a series of gradual changes. In the case of cells this is of course nonsense. The changes needed occur automatically and quickly in a developing embryo. Since all organisms are composed of cells it seems rather strange that changes in fully developed organisms are considered from an evolutionary point of view, rather than as just an extension of what occurs in the developing embryo. Since what causes the development of cells into different kinds of cell is not really understood, how then in evolution can those changes of design, which are even more dramatic, produce the different organs which are required to meet the demands of moving from, for example, an aquatic to a terrestrial environment, be explained? Darwinists would insist that it is just chance factors followed by natural selection. There is no reason to suggest why this should be so, and it is simply used because we do not know as yet how it occurs. It cannot be just chance that a nerve cell or a muscle cell arises as the cells divide from the original zygote. The original fertilised egg knows exactly what is expected of it, a chicken is a chicken from the word go!

Darwinists argue that for a new feature to evolve it has to be advantageous. If indeed it is then might it in fact be just the result of not a chance mutation, but part of some original plan waiting in the wings for the right moment to make an appearance? In the evolution of the limbs from fins, and the subsequent change of legs into wings, the pentadactyl limb is cited as evidence of these changes. Both limbs and wings have the same 1-2-5 bone formation. This in itself has no obvious advantage and has to undergo a great deal of modification in order to produce a functional organ. The pentadactyl limb however arose in the amphibia from the fins of fish, providing the basic pattern from which all future modification could be based. Therefore the original pentadactyl limb seems to have been designed with such future modification in view. Was this sequence chance, or was the evolutionary line already laid out? Evolutionary changes are very efficient and like any journey preplanning will increase the efficiency, as will a distinct goal to aim for. Evolution may have taken the course it has as a lucky result from wandering in the wilderness, but somehow I am not convinced. I have suggested that the patterns of evolution appear to follow clear cut lines and sequences of events which would point to a predesigned plan. The question which arises is that if this is so then where could the plan be? In this I would support any of the evolutionists in the search for evidence. It may not produce conclusive proof, but it may produce a better hypothesis than the one we have at present. One line which is worth considering is the role of the ninety percent of genetic coding which appears to have no known purpose.

So far I have considered some of the factors which are put forward as being responsible for evolution. There can be little argument that variation resulting from the combination of genes as a result of sexual reproduction does provide an advantage for survival. This however cannot be responsible for the major changes which would need to have taken place to produce the main classes. Mutations do occur in the normal process of cell division as either accidental mistakes in the copying process, or as a result of external factors such as those arising from the effects of radiation or toxic chemicals. Although there is no evidence that the incidental effects of these have ever resulted in any that were not detrimental to the organism, it appears that this is the only way these major changes can be accounted for by present-day theories. Whatever effects were needed to result in the change from fish to amphibia, and reptiles to birds and mammals, they had to be achieved quickly. Gradual changes over a long period of time would I contend be a disadvantage and in the general concept of survival of the fittest would die out. Missing links are on the whole missing. Any fossil which shows the slightest possibility of filling one of these categories is grasped at by the Darwinists as a means of supplying the evidence needed to show that this is what occurred. I contend that if these changes were indeed gradual then there would be no shortage of fossil evidence.

I have also considered the differentiation of a single cell. Life starts from a single cell, which then divides and develops into the full multicellular organism. This means of course that every cell carries the complete plan of the fully developed organism. The cells differentiate quickly to fulfil their predetermined role in the production of the various tissues, be it nerve, skin or muscle. This process is not the result of chance. It is a built-in programme. Why then, I ask, should the change of one organism into another not be the result of a similar programme rather than just the consequence of chance mutations?

Chapter VI
LAMARCK: A THEORY OF EVOLUTION

Jean Baptiste de Monet Lamarck looked at evolution from a slightly different viewpoint to that of Darwin. He published his theory in 1809, the same year that Darwin was born. Although his ideas have been refuted, his was the first attempt to explain the origin of species by evolution. He was also the first to realise the significance of fossils, and their relationship to living organisms. He reached his conclusions from observation of what he witnessed in nature, and his explanation has caused a great deal of controversy over the years. Let us examine some of his conclusions. He proposed that evolution arose because of the following facts. Organisms starting from simple microscopic forms tend to increase in size and complexity. Organisms arise because of certain needs. The use of an organ causes it to develop further, while disuse causes it to degenerate. The idea, however, which created the most controversy was that changes in an individual during its life are inherited by its progeny. Darwin too believed that this was true but did not press the point as intensely as Lamarck. Lamarck assured his place in history by his association with the term 'inherited characteristics'. It was this theory which created the greatest trouble. When one considers why this was so, one reaches the conclusion that it was simply because it doesn't seem to make sense. To be fair this seems a rather poor approach, since very few of the accepted theories make much sense, and this does not necessarily mean that they are not true. It is also argued that no experiment has ever been carried out which proves that acquired changes in organisms can be passed on. One has to respond by pointing out that no experiment has ever shown how life could have evolved.

Experiments were set up to test his theory. Newborn mice had their tails removed at birth and then allowed to interbreed. Should their offspring be born without tails, then Lamarck's theory would be acceptable. After many generations the offspring were still being born with tails. This is hardly surprising. Let us assume that Lamarck was correct and that inherited characteristics play a crucial part in evolution. Then the inherited characteristics would have to be beneficial to be inherited, and that would rule out being born without a tail; also the loss of something is hardly an acquired characteristic but the reverse. Lamarck also noted that if an organ was not used then it would degenerate. There is evidence that this does indeed occur, and I will be returning to this later. The point is that this is just as difficult to explain as the rest of his theory. I am not saying that Lamarck was necessarily right, but I am saying that the accepted evolutionary theory does not explain a great deal of what Lamarck at least attempted to explain.

Lamarck suggests that the body changes as a result of the environment and its effect upon the organism, and that these changes are passed on to its offspring. I accept that this appears unlikely. It is rather like a builder building houses to a plan which specifies four bedrooms. The people who buy them only ever use three of them, but unless the builder can be informed, and his plans changed, he will continue to build four-bedroom houses. In living organisms it would make sense that the plans could be changed to accommodate the needs of the individual during its life. The argument against this is that the genetic code cannot be changed by the actions of the individual, and therefore cannot be passed on. What is known is that organs which are used regularly do slowly change, increasing in their efficiency and in some cases their size, while others will deteriorate when they are not used. Muscles are examples of this, and the body responds to their use by increasing their size and efficiency. The genes have to be affected by this use or lack of use in order to respond to their need of increased or decreased protein production, and in the case of muscle cells, by an increase in the number of mitochondria in order to increase energy production. Equally the genes take into account the lack of use of organs by reducing the production of materials to maintain them. How are genes made aware of this in order to instigate these changes? The whole organism is affected by this factor. When we cease using muscles they deteriorate; why should this be so? The organs, and therefore the tissues, and therefore the cells, and therefore the genes making up the DNA, respond, and therefore are aware of what is happening.

People who live at high altitudes have a higher level of haemoglobin in their blood in order to compensate for the low oxygen level than people living at low altitudes. One could say that this was the result of natural selection and the survival of the fittest. This however is not the case. Athletes in the Mexico Olympics found that if they lived at the high altitude for about a week then their haemoglobin levels had increased. When they came down to compete at low altitude their performance was enhanced as a result of this increased oxygen capacity of the blood. After about a week at low altitude their haemoglobin level was back to normal, and so were their athletic performances. In other words their bodies were capable of responding to the environment.

Lamarck said that these kinds of responses could be passed on, and whether it is true or not, it is a far more convincing solution than mutations and natural selection. At least the organism is responding positively to the demands of the environment, rather than waiting and hoping that by some miracle a mutation will arise which will provide the response. The main objection to Lamarckism is that the sex cells are separated from the body cells, and hence any bodily changes which occur during the life of the organism cannot be passed on to its offspring. In other words, the hereditary material is not aware of what is occurring in the body and therefore will not change. Surely, however, for organs to reduce in size and efficiency when they are no longer needed, and for this to become a feature of the progeny, such as the loss of functional eyes in moles, or the reduction of the size of the appendix in humans, shows that there is a response in the genes, and they are changed. How could this occur? Some biologists have suggested that information may flow back into the DNA and that the genes can be changed. Viruses could be used to carry the information from the body cells to the DNA in the sex cells. It is easy to ridicule such ideas by saying that there is no evidence to support them, but in fairness there is very little evidence for many of the accepted theories concerning living organisms. Animals can and do build up an immunity to certain diseases and also to toxins. The question is, can this immunity be passed on to the next generation? The answer would appear to be no, but since the organisms responsible for the diseases and the production of toxins are also changing we cannot be sure.

Both camels and warthogs spend a great deal of their time with their front legs bent in a kneeling position. The parts of their legs which are in contact with the ground have a pad of thick skin and hair which acts as a cushion on the hard ground. These pads are present from birth. Assuming that Lamarck was wrong and they could not have originated from the use of these parts of the legs for kneeling, then we are left with the fact that they are the result of a mutation which provided the animals with protective pads and this gave some form of advantage over other members of the species who did not have them. As a result of this all camels and warthogs have evolved this protection. It seems rather strange that the protective pads only occur where they are needed and nowhere else on the body where the possession of such protection would be of no disadvantage to the animal. The soles of our own feet have thick layers of skin which provides them with protection. Again this begs the question, did this arise by chance or, what makes more sense, did they gradually develop to provide a functional protection? The idea of acquired characteristics being passed on does make sense; it is not however accepted simply because no one as yet can explain how this could be achieved.

There does exist some way in which cells are able to know and respond to situations occurring away from themselves. Should a limb, or in certain cases just part of a limb, be removed, it will often be fully replaced. Regeneration of limbs does not occur in more advanced animals, although the repair of injured parts does. How are the cells aware of this damage and how do they know when the repair has been completed? I make no apology for posing these questions. Science must ask questions whether or not they can be answered in the context of our present knowledge. What tends to happen is that they are put to one side so that present theories can remain intact.

Let us return to what Lamarck was saying, namely that acquired characteristics are inherited. It is certainly difficult to see how certain behaviour can be acquired in any other way. Behaviour is learnt by experience, and if this experience is not passed on then each generation would have to learn from scratch. Certainly in higher animals the answer is that they can be acquired by copying, or by communication; in lower forms of life this is not the case. Certain insects are brightly coloured and poisonous. The colour denotes a danger to birds and thus they avoid being eaten. Does this mean that every bird has by trial and error to have to learn this fact, or is this information present from birth? It must be remembered that all information which results in a change in behaviour was new at some time and it then becomes incorporated into the normal behaviour of the species. The complicated methods of courtship in certain insects and spiders, the roles of individual insects living in colonies, ants and bees etc. are inherited characteristics. It is all very well saying that this is instinctive behaviour, but it had to arise and be stored somewhere. The behaviour patterns are uniform throughout the species, and therefore the information must be the result of the transfer of genetic information. Each organism is starting from one cell and one set of genes. The behaviour therefore has to be retained in that genetic material to produce what we call instinctive behaviour.

I have looked at the way in which certain characteristics may be passed on from one generation to another, and I have shown that Lamarck has become a name which has no place in our present evolutionary theory. The reason is that although his concept of the way in which certain characteristics are acquired would make sense, they cannot in the context of our present understanding be explained. Even if we accept the evidence for the origin of certain inherited physical attributes, behaviour I submit still presents a problem. Much of behaviour is the acquired result of experience: and yet much of this behaviour is present from birth. Since each organism starts life as one cell, then this information for the production of such behaviour has to have been present in the genetic information in that original cell from which the organism developed. The evolutionists accept this, but explain that it is the result of the gradual accumulation of beneficial changes which being advantageous were passed on through the species by natural selection. After a long period of time this behaviour became instinctive, and thus became the norm for that particular species.

Let us consider the behaviour of social insects. In the case of ants, bees, wasps, termites etc., each member has a defined role in the colony. Their activities are instinctive and are therefore the result of the gradual accumulation of chance changes in what would have been normal individual and competitive behaviour. A worker bee just happens to have acquired the ability that on finding a food supply it would return to the colony, and by a complex dance demonstrates the locality of the food supply to the other bees. These in turn just happen to have acquired, by a series of chance changes in their genes, the ability to understand the meaning of the dance, and so are able to locate the food. Let us be clear, this type of sign language is complex. Two kinds of dance are performed, the round and the wagtail. The former is used when the food is close, the latter when it is at a distance of further than one hundred metres. The dance is complicated. By using a variation of the speed, the length of time the dance takes, and other factors, a directional bearing is given relating to the angle of the sun. Now I cannot accept that this behaviour could be the result of gradual changes which because they had some survival advantage were retained and passed on. It had no advantage to the colony unless it worked from the beginning. As my son is apt to say in response to some outlandish story. 'If you believe that you will believe anything.'

Much of behaviour is simply learnt by copying other members of the species, but this of course, if one rejects Lamarckism, cannot be passed on. Even if one accepts this as true, how can one explain the behaviour of, for example, a cuckoo? It starts life as one cell in an egg which has been laid in the nest of another species. Once hatched it is reared by them and then come autumn it will fly from the British Isles to Africa for the winter. First year cuckoos delay their migration in order to increase their strength, so they leave after their parents, thus making the journey on their own. The cuckoo returns in the spring. It recognises the call and appearance of other cuckoos, which is essential in order to find a mate. This is surprising, however, when one considers the lack of any imprinting after it was born, which might be considered essential in order to relate to its parents, and thus create its own identity. After mating, the female locates a nest and lays one egg; the egg she lays tends to be the same colour as the eggs of the species whose nest has been selected. The laying of a single egg is repeated in other nests. The point I am making is that none of this behaviour has been learnt from its parents. The whole of its behaviour has therefore to be present in the genes, in the gametes or sex cells. How did it get there? All the result of chance mutations? Lamarck's theory of acquired characteristics may not be explainable, but I have to say that very little of our present theory offers any convincing alternative explanation.

Chapter VII
THE EVOLUTION OF THE MAIN CLASSES

So far the evidence for natural selection as a means of determining the small changes in living organisms seems to be acceptable. The evidence from straightforward observations of present-day organisms, both wild and domesticated, confirms that this is not only possible, but does in fact occur. However, the evidence that the combination of resulting mutations and the process of natural selection could have led to the great changes which would have been required to produce the major classes is not exactly convincing. It is this evidence which is important. Without it any explanation of the origin of the living world becomes mere supposition, and evolution without evidence has no greater authority of truth than the creationist theory it attempts to replace.

I suggest that there is no real evidence to show how the major classes evolved. For gradual change to have taken place there would need to have been a great many intermediary links. The alternative to this was a big jump, which even the great champion of Darwinism, Richard Dawkins, acknowledges as being highly improbable. He supports the idea that it could be accomplished by a series of steps. There is unfortunately a surprising lack of evidence of these steps, and that basically is the problem. The missing links are indeed missing. It is true that if one examines for example the invertebrate fossil record, one finds that there is a great deal of fossilised material. This shows that even over long periods of time no great changes took place. It is true that there are examples of unique forms appearing, but these do indeed appear to have been unique, having no connection with existing forms, and did in fact disappear from the scene quite quickly. There is unfortunately no fossil record connecting any links between the major classes. It is apparent that when they do appear they already have the main characteristics of their class. The earlier rocks preceding those that are rich in their fossils, and in which one could reasonably expect to find the forerunners of these animals are virtually barren. Why are these rocks basically empty of fossils?

The same kind of argument can be applied today. The difference between similar animals, dog and fox, mouse and rat etc. are all rather trivial. Yet if one looks at the differences between a crustacean and an insect, a fish and an amphibian, or a land mammal and a whale, then tremendous changes would need to have taken place. Where are the transitional stages? The answer appears to be that there are none.

For mammals to have evolved gradually, there would be no gap between mammals and the rest of the animal kingdom. So far there is no candidate for the ancestor of the mammal. It is suggested that mammals evolved from a mammalian type of dinosaur, but that they were not successful until after the dinosaurs became extinct. Surely, in this case, the whole essence of evolution, natural selection, survival of the fittest etc. must be placed in question. There was obviously no advantage to be gained from the whole range of changes required for the mammal to arise; warm blooded, hair, internal development of the young, mammalian milk etc. All of these and more to produce a small shrew-like animal that had no advantage over the reptilian dinosaurs. Did all this occur on the off-chance that the dinosaurs might become extinct? It is argued that because they were small these mammals could not compete. Small mammals today, however, have proved to be very successful, and in any case why should the small mammals have not increased in size?

If we accept that the mammal's origin lies in a form of mammalian dinosaur, then how is it that in geographically isolated populations it went on to produce similar groups of mammals from this tiny shrew-like ancestor? The evidence provided for the evolution of mammals has to be open to question. The suggestion that the reptile changed, and that the resulting mammal-like reptile provides the answer may be well off-beam. The fossilised candidate in question may well have been completely reptilian. There is of course very little evidence of the internal structure of the organs in the fossil record, and the assumption is based entirely upon the shape of the skull and jaw. This, I contend, is a big assumption, and certainly if we were to use such flimsy evidence to classify present-day animals it would be laughable. Convergence occurs in all living creatures, and one might just as well say that bats evolved from birds because both can fly and are adapted to eat insects. It is worth noting in passing that most insects have evolved wings in order to fly, and so have followed when they evolved from crustaceans the same pattern as birds did when they supposedly evolved from reptiles. Quite a coincidence!

Mammals did of course arise, otherwise we would not be here today. For them to have done so in the way suggested makes little sense. A tremendous amount of change was required to produce a mammal from a dinosaur, and after having produced all those changes the mammals still had no advantage. For the mammal to have evolved from the dinosaur, then each gradual change should have proved advantageous, thus triggering the next stage of evolution. We are told however that the finished article was disadvantaged, so why did these changes occur? During all of this time the dinosaurs just evolved different ways of exploiting all available resources. However, once they had become extinct the mammals were set to evolve, resulting in the diversity which we see today. It is surprising that the dinosaurs' evolution produced by comparison rather minor changes, and the extinction of the dinosaurs does in itself present something of a problem. The general opinion is that it was brought about by the impact of a meteorite. From the fossil record, we know that it took a long time for the extinction to occur; it was probably due to the aftermath of such an impact and not just the impact itself. What is of interest is that the mammals and birds managed to survive, but the dinosaurs failed to do so. For a class that had such domination of the earth on land, at sea, and in the air, this seems rather strange. One cannot help but wonder if some other factors were responsible. However, with the evolution of the mammals something special had to occur to result in the gap between the primates and other mammals.

The evolution of the horse is frequently used to demonstrate the way the different changes occurred from the early to the modern horse. These differences are, compared to the changes needed to produce the mammal from the reptile, very superficial, and yet it took sixty million years for the horse to evolve from its early forms to its modern equivalent, the process taking a separation of ten genera to accomplish. Think of the number of stages which would be needed for a simple mollusc to evolve into an arthropod.

When animals that have become extinct are used as evidence for missing links then one must be wary of such claims. The coelacanth was once thought to have been such a bridge between the fish and the amphibia. Unfortunately for theorists it was discovered alive, and it was subsequently demonstrated that there was no evidence of it having organs pre-adapted for use on land.

If the other theory was correct, that changes occurred suddenly as a result of macro mutation, then there should be countless numbers of failed monsters in the fossil record in order for evolution to have produced some animals with changes which were advantageous. It must be remembered that most mutations have harmful results, and therefore fail to effect evolution. The fossil record shows no failed monsters, and therefore one has to conclude that the idea of non-selective mutations as a method for producing great changes must be discarded.

The idea of sudden change was taken up by Eldridge and Gould, who produced the punctuated equilibrium theory. At least their belief that some other explanation was necessary in order to explain the lack of intermediary types is encouraging. Their idea was that evolution occurs in fits and starts, with long periods of stasis. Species, they suggest, arose rapidly in small isolated populations and then spread over a wide geographic area, undergoing little further change. This would account for the gap between species, but it does little to account for the large gaps between the classes, fish–amphibia, reptile–mammal etc. Representations of all the main classes are fully developed members of their class when they first appear in the fossil record.

It can be argued that the effect of change does not always arise as a direct result of the change. It might be said for example that feathers arose for flight rather than as a means of temperature control. This type of view does not however take into account the other changes needed for flight, the movement and development of forelimbs into wings, the change in the leg joint for perching, and the changes in the structure of the bones for a reduction in weight. The Archaeopteryx, fossil of a reptile-bird, is used to demonstrate the interim stage. The point is however that it had feathers and therefore could presumably fly. Recent research suggests that powerful flight would in fact have been possible. The fossil admittedly hints at reptilian ancestry but there would have had to be a great number of intermediate stages. The feathers might merely have been the result of convergence, while the soft internal features may well have been totally reptilian. It is difficult to see how scales could gradually convert to feathers, or by any similar gradual process result in the production of the avian lung.

The critical thing about fossil evidence is that the soft parts are missing. One has only to consider the human skeleton without prior knowledge of living human anatomy. Ears, nose and breasts, and male genitalia, would be absent and therefore would not exist. The significance of the absence of soft anatomy is given by the example in *Evolution A Theory in Crisis* by M Denton. 'Suppose all marsupials were extinct. From fossil records we would assume that their reproductive biology was no different from that of mammals.'

Even the assumption that the mammal-like reptile in which the skull and jaw were similar to the mammalian conditions may be erroneous. Lack of internal evidence means that it may have been entirely reptilian. The cranial endocast shows that the animal had a typical lower vertebrate size which in no way approached mammalian size. So the evidence is anything but conclusive and merely depends upon how it is used to support a preconceived idea.

The pentadactyl limb, which is the formula for all terrestrial vertebrate limbs to be constructed from the same basic design, supposedly provides evidence that all such limbs are derived from a common ancestral source. The difficulty is that all hindlimbs have the same pentadactyl pattern, yet there is no claim that they evolved from the forelimbs or that both the hindlimbs and forelimbs evolved from a common source. Why, if the amphibia had evolved from fish, are the front and rear limbs the same when their origins from the fish were not the same? More questions, but few answers. Answers are only valid if they are correct, otherwise they are no more than guesswork.

So far I have examined the evidence, or lack of it, to account for the major step in evolution. The greatest challenge to our present theory of evolution has to be the explanation for the origin of the main classes of living organisms. Each of these classes appear to have arisen spontaneously with very sparse evidence to explain how the great changes which were required occurred. Unlike the small differences in species, dramatic changes had to have taken place, not only for the changes in the physical structure of the bodies but also in every system needed to sustain life. There had to be a change from cold blooded to warm blooded animals, and changes in the methods of reproduction, as well as changes in the methods of locomotion, digestion, respiration and excretion. There is no logic in arguing that these changes could have been achieved by a gradual process because any attempt to do so would have been against the main principle of evolution, i.e. that any chance changes which prove beneficial even to a small degree will be retained and passed on. Changes occurring in this way must initially place the organism at a disadvantage. Putting it simply, if any of my limbs showed signs of evolving into flippers or wings, then I doubt if I would survive long enough to pass this on to any future generation so that they would be able to continue the process. Equally any such change would hardly improve my chance of competing with the rest of my species for the opportunity to do so. This may seem flippant – no pun intended – but this is what we are expected to accept, not just once, but to account for how all of the main classes arose. There is no hard evidence that this is what occurred, but because there is at present no other acceptable explanation, then it seems that this has to be accepted as the way it happened. Where, I ask, are the missing links? There do not appear to be any, despite the efforts to produce controversial evidence to the contrary. In the same way that a ship did not gradually change into a train, or a train into a motor car, or a motor car into an aeroplane, so the major classes could not have arisen by a series of gradual changes. Once the main changes had taken place then I agree it is possible that gradual changes over a period of time could have occurred to suit the different roles of the species which make up the classes. Man's origin as a branch-off from the apes created great problems for the acceptance of Darwinism, but this is far more plausible than the production of birds and mammals from reptiles. There are indeed problems with the evolution of man, which will be dealt with later, but this is merely in keeping with the rest of the problems faced by our present evolutionary theory.

Chapter VIII
PROBLEMS?

Let us now consider some of the facts concerning the theory of evolution. To begin with, the present theory of evolution cannot be proved by the methodology used for the normal acceptance of a scientific theory. However this should not be taken to present too great a problem. Biology by its very nature has never been an exact science like mathematics. The biochemist may tend to disagree with this and the geneticists would almost certainly disagree. This is understandable, because they are attempting to explain biological concepts using the same tools one would use to attempt to understand how a machine operates. It would be interesting to see how the same method could be applied to determine the function and purpose of a piece of sculpture, or a painting. Both the machine and the works of art are products of the same biological organism. Biology may have its foundation in the world of the physical sciences, but it has evolved into a freedom of action which cannot always be transferred into mathematical equations and predictability. When it appears to do so then it is grasped for eagerly by the scientist, who feels reassured to be back in a world where one and one always makes two. This creates an insulation against the time when the answer happens to come to three. Then it has to be ignored so that it will play no part in the mathematical precision of a new theory.

However, my point of view is that if one studies for example behaviour and variation in the nature of animals, then one immediately finds difficulty in attempting to explain these in the accepted way of cause and effect. The fashion today is to explain all types and variants in the context of their genetic make-up. A criminal, a homosexual, an alcoholic all must result from having a genetic origin which can account for this variation from the accepted norm. It is easy to fall into this trap. Hair and eye colour, body shape etc. are genetic characteristics which can be explained by a certain genetic make-up. To extend this to explain the whole of human behaviour is I suggest naive. The counter-argument would be that perhaps not all behaviour, but most, is due to genetic make-up. I am sorry, but it has to be all or nothing, otherwise we are left with a residue of individual likes and dislikes, appreciation of the arts or indifference to them, and the whole range of individual characteristics which cannot be explained by the present genetic theory. What gene, I wonder, is present in my genetic make-up which is responsible for my questioning of the present evolutionary theory? Nurture of course is responsible for certain behaviour, or so we are led to believe, and nature versus nurture has been an ongoing debate for a long time. The difficulty is that unless all behaviour is the result of nurture, then how does any other behaviour become incorporated into the genetic make-up of the individual? This I contend is the problem. The theory of evolution accepts all that fits and ignores all the facts that don't. It is a dangerous way to do science. One starts with a theory and collects all the facts which support it, and ignores all the facts that don't. The other method is to look at all the facts with no preconceived ideas, and then look for a theory which would explain all the evidence. The attractiveness of the first method is easy to see, as it guarantees a theory. The second method can leave one completely baffled.

Darwin himself, in fairness, employed the second method. He collected the evidence, and the theory came afterwards. He himself was conscious of the criticism, and probably had some doubts as regards the theory. Its acceptance by the scientific community was influenced by the fact that it was the only alternative to the creationists' view. Although Darwin's work is titled *The Origin of Species*, he does not in fact claim that his theory leads to the production of new species, simply variation due to natural selection. Natural selection does of course work and has been confirmed by the studies of birds, insects and snails which have been geographically isolated on islands and have diverged due to natural selection. This of course was what inspired Darwin in the first place, but to extend this to account for the production of the main classes of living organisms places the ball in a different court. Here the evidence becomes extremely slender. Darwin was the first to acknowledge the lack of evidence to be found in the fossil record. The Darwinian view, and this should not be confused with Darwin's own view, sees change as inevitable. This however confuses variation with evolution. The result is not the same. Variation and its effects are reversible, but the changes which are needed to produce a new class of organism are not reversible. A fin may evolve into a limb but it does not reverse back into a fin. A foreleg may evolve into a wing but not the reverse. The difficulty with the concept of gradual change is the lack of evidence. As has been stated, the fossil record shows not gradual change, but a resistance to it over long periods of time. New groups suddenly appear with no evidence of intermediary stages. This is what one might expect, even though it is difficult to explain. A lizard functions as a lizard, and a bird functions as a bird, but anything in between would lose out. This fact was noted by the eminent biologist, the late Stephen Jay Gould. He argues that the fossil record does not support the idea that evolution is a process of gradual change. Instead the fossil record supports the idea that new forms suddenly appear, followed by long periods of 'stasis' when little change occurs.

It is generally considered that the criticism of Darwin is due to the fact that he was unaware of the function of genetics, and in particular the part played by mutations. There is however a large gap between the understanding of the genotype, the genetic make-up of the organism, and the phenotype, its visible characteristics. Natural selection acts upon the phenotype but the results are the products of the genotype. The relationship between the two in the evolutionary theory is difficult to understand. The arising of a mutant gene which would bestow some beneficial effect upon the phenotype is hardly likely to result in any significant change.

Darwin was of course unaware of the part played by genes, and even as far back as 1867 F Jenkins wrote 'It is impossible that any sort of accidental variation of a single individual, however favourable to life, should be preserved and transmitted by natural selection. The advantage whatever it may be, is utterly outbalanced by numerical inferiority, and (such) variation would be swamped.'

The discovery of the role played by genes shows in fact that indeed it would not be swamped, and could exert an influence upon the population. That is fair enough, but can micro-evolution evolve into macro-evolution? In demonstrating the evolution of any group, invertebrates or vertebrates, the practice is to show the gradual change of organs over a long period of time. The trouble is that this does not always follow. What does occur is that the environment also changes from, for example, aquatic to semi-aquatic to terrestrial. In other words, it could be that it is in fact the environment that is dictating the changes, and if the environment does not change then neither does the organism.

Where there was a sudden change in the environment which was too great for organisms to survive, then there would have been no subsequent generations. Survival then would have been a matter of chance, and once the catastrophe was over there would be less competition, and empty niches waiting to be filled. This would of course account for the lack of fossil evidence showing a gradual change from one form to another, and also account for the long periods of stability. However it would also have had to either cause the mutation to occur quite dramatically in order for there to be some organisms where it would prove to be advantageous, or there would have to be some way in which the mutation could remain dormant within the genetic code until such time as environmental changes would trigger its appearance at a time when it would be most advantageous. The arising of new forms does seem to have occurred quite readily, especially in the invertebrate world. The fossils of the Burgess Shale, a fossil area in British Columbia, showed quite clearly new and unique forms of invertebrates. Ten completely new invertebrate phyla were discovered, but significantly none were links between previously known phyla. It can be argued that without intermediate forms then evolution cannot be taken seriously. We are not just concerned with fossils but also with the possibility of living missing links.

Although I have concentrated on animal evolution, it must be noted that the same defects in the present evolutionary doctrine apply equally to the plant kingdom. The sequence of plant forms from mosses to ferns, to conifers and to flowering plants shows clear cut divisions between the classes. There is no evidence of one evolving from the other through a series of stages.

It would seem that in order to come to grips with the problems of evolution one should look at the evolution of the building blocks of life itself, the cells. When we understand how these are capable of evolving and differentiating, then it may well go some way in explaining how multicellular organisms are able to do so. The differentiating of cells from a single zygote poses the first question. Where does the control come from? It would appear that the position of the cells determines their development. For example the colour of the ear tips of certain animals are often darker due to the temperature, where the development of a pale colour needs a higher temperature in order for it to be synthesised. It would also appear that certain genes are switched on when certain chemicals are present. These chemicals have however to be manufactured in order for them to function as switches, and this manufacture must be gene-controlled. The genes then do seem under certain conditions to be able to respond to the environment, and presumably evolved to be able to do so. We have the hint of Lamarckism raising its head once again. When we return to the topic of repairs to the body, we find ourselves asking how can genes, which it is assumed do not change their role, yet in the case of regeneration, do just that? It could be argued that the act of damage stimulates the chemicals to be produced to trigger off the genetic response to regenerate. How do they know when it has completed the task? One point of interest here is that invariably regeneration of any lost part results in a replacement, but one which is smaller in size than the original; why?

There is also of course the problem that there are no intermediary stages in cells, for example nerve cells, or muscle cells. Each cell's role is clearly defined after its subsequent divisions from the zygote, and yet since it carries the same genetic instructions it must have the potential to develop into any kind of cell. This is what we see in the main animal classes; no intermediary stages. One may consider that the stages between cells don't occur because of certain constraints or limiting influences. Forces, chemical or physical laws, dictate what can or cannot occur. Should this be true, then it would mean that the environment is not the sole factor in deciding the survival of the fittest.

Should I wish to bake a cake, I would need certain ingredients. I would of course have to know in advance what these ingredients were, in other words I would have to know the recipe. I then have to know how to combine them in the correct order to produce the cake. As any cook knows, the result can never be totally guaranteed. The cell does not have the luxury of accepted failure, and has to get it right every time, without presumably knowing what the cake is for, or when it is ready to eat. One has to conclude that there is more to it than is at present suggested. Multiplying these factors in multicellular organisms and one can appreciate the concern of many scientists. In the book *The Eighth Day of Creation* by Horace Freeland Judson, the author quotes Francis Crick, who having jointly discovered the double helix structure of DNA now expresses his concern with the problem. 'Although a great deal is known about the metabolic reactions which take place within a single cell, little is known about how this is achieved in multicellular organisms. Here one faces problems of development, embryology, differentiation of the different kinds of cells, organs, tissues, the healing of wounds and regeneration.'

While we accept that a single cell (zygote) can evolve into any of the specialised cells making up a multicellular organism, is it not therefore possible that it also has the function of carrying the code to produce the changes required for the evolution of the organism? In other words, the cells have the ability to determine the course of evolution. In answer to the question posed of where is the information for this to happen, one must remember that most of the genetic code, junk DNA, appears to have no purpose. Perhaps this is its actual function.

We now turn to a subject which seems to be largely ignored in the context of evolution and yet one which is critical unless it can be explained within the evolutionary theory. Whether a species survives or not depends more upon its behaviour than anything else. Frequently the modification of structure comes after behaviour. Sea mammals took to the water before having the adaptive structures to do so. Should this be correct then behaviour could be the instigator of evolution, as opposed to the result of it. The problem is that the genetic code does not show any signs of being able to respond to specific behaviour patterns. Complex reflex behaviour such as nest-building, the construction of a web by a spider etc., cannot be accounted for by the mere production of certain proteins. Hormones obviously trigger off the desire to build a nest, but the ability to do so has to have been passed on from one generation to the next. Presumably the nest may have been primitive in its original form, and by trial and error efficient nest-building was achieved. At the time of writing, two pairs of birds are nesting in my garden, a pair of blackbirds, whose nest is lined with dried grass, and a pair of song thrushes, whose nest is lined with mud. The difference in the two types of nest is as fixed as the colour of the two birds' plumage. The latter can be explained by the genetic information, but what of the nest-building? Is this information incorporated into their genes, and if so how did it get there, and how is the information translated into the complicated behaviour needed to build the nest? As the critics of Lamarck pointed out, learned behaviour cannot form part of the genes and be passed on, because we cannot think of any way this could happen. Well, it seems that complicated reflex behaviour certainly can be passed on, and had to have originated as the result of learning from experience. How, however, could this experience have been passed on and incorporated into the genetic code, and after doing so become the instigator for the next generation to be able to repeat the task? The information would have to be incorporated into the genes within the sex cells. This point is taken up by Rupert Sheldrake when he asks 'How does the inheritance of certain proteins make swallows, for example, migrate from parts of England to South Africa and then return?' Sheldrake in his endeavour to answer this and other questions proposes a field theory, which suggests that force fields are able to influence other fields in space and time. He receives little encouragement from the rest of the scientific world and admittedly has no proof to back his theory. All he has are observable facts, which, as no one else was prepared to try and explain, he offered a possible line of inquiry. In the book *The*

Descent Of Darwin by B Leith (1982) the question is asked 'Why must Darwinists try to open all the doors of nature with the same key? Is it not reasonable to assume we need more than one? How can biologists ask fresh questions about evolution if the determinists keep insisting they have all the answers?'

Darwin was not dogmatic. He was ready to accept new perspectives, but his disciples have ignored this. The pathway of development from gene to the total developed organism still remains a mystery.

So far I have considered the problems faced when there is an attempt to produce hard fact theories in biology using the same rigid methodology which is employed in the study of the physical sciences. Sociology, which is classed as a science, suffers from the same difficulty. It is of course concerned with the scientific study of accumulative behaviour, and its explanation, in a biological species, i.e. human beings. There has been a recent trend to combine both biology and sociology into one subject. Since sociological factors must have their origin in the biological make-up of the individual, then the explanation of how societies function has its origin in the evolution of the individual. The point being made is that in the study of any group behaviour within a species, one has to be aware that the effect is influenced by the behaviour of the individual organisms. This is the same as the way in which an individual organism is the product of the cumulative effect of the cells of which it is made.

Sociobiology is now studied at several universities. Investigating the link between the behaviour of living subjects and the cause and effect factors is at least attempting to get away from the restraints of theories which have to fit into the rigidity of accepted scientific principles. An example of a case where what appears to be a perfectly logical explanation can easily be overthrown is demonstrated by a study of suicides by Emile Durkheim. Suicides are generally considered to be the ultimate result of an individual's inability to cope with certain circumstances and social pressures. It could reasonably be expected that in times of social upheaval such as war, mass unemployment and extreme poverty then the number of suicides would rise. Durkheim showed that suicide rates do in actual fact increase dramatically during times of social stability, and fall during times of social upheaval. As far as I am aware Durkheim's findings have yet to be explained. This example demonstrates the danger of jumping to conclusions, however obvious they may at first appear.

Returning to the topic in hand. The danger of attributing all observable biological evolution to the chance mutations of genes is that it can blind one to the possibility of the involvement of other factors. I have considered the different roles of the genotype and the phenotype, and the problem of dealing with the long periods of time when there is little evolutionary change, and how this can then be followed by a sudden explosion of new species. I have also considered the fact of each cell having the potential within itself to develop into any type of cell and therefore logically into the complete organism. This would therefore suggest that somewhere in that system lies the instruction to instigate evolutionary changes. Cells do not change their role and structure by a series of chance changes. They respond when it is required for them to do so. Every cell carries the ability to reproduce and this ability is carried forward into the whole organism. Here is a reflection of the way the behaviour of individuals results in the way societies behave. I have considered how the sex drive, without which there would be no evolution, is present in all living organisms, an extension of what is present in each and every cell.

Sexual reproduction resulting from the sexual drive has become a priority in living organisms. More than that, it appears to be the purpose for their existence. Without it of course evolution would not have occurred. Why should this be so if the whole process of evolution is based upon chance changes being advantageous to the individual? Why is the whole process of reproduction so important? It is beneficial for the process of evolution but not to the individual. Death, which removes the instigators of reproduction, occurs when the individual's genes have been passed on and sufficient time has elapsed for them to assist in the care of the progeny.

Why has this ageing process evolved? Once again it is certainly not for the individual's benefit. This seems to point to the fact that far from being a chance process, the purpose of life has always been for it to evolve, aiming for some future predetermined goal.

I have also examined the problem of explaining the origin of complex reflex behaviour. This is passed on to each generation, but for this to happen then its origin has to have been incorporated at some time into the genes. Failing this then it was present from the beginning. The question can also be raised as to how the mind and body relate. Is the mind a product of the body or is it the result of the process of learning? Certainly there is evidence that the mind can influence the activities of the body. I make no pretence of knowing the answers, but these questions must be asked and not just set aside. They must be in the front line of future research, which must be carried out with an open mind resistant to scientific bigotry.

One type of behaviour which has presented problems over the years is altruism. Although it seems quite reasonable to accept that genetic mutation arising by chance and producing factors which result in competitive and therefore a survival advantage to an individual organism, will be retained on the basis of survival of the fittest. There are however certain factors which appear to work against this principle. Certain instinctive behaviour, which may arise due to the presence of specific genes, results in beneficial factors to others while reducing the survival chances of the individual concerned. This is referred to as altruism. An organism is said to be altruistic if it increases another organism's chance of survival and therefore reproduction at the expense of its own. There are several theories which have been formulated to account for this. Richard Dawkins refers to this as the selfish gene factor. Here the selfishness means ensuring that copies of an organism's genes survive, and it does not matter in which organism they survive.

A bird may produce an alarm call when danger threatens, thus placing itself at risk, while its action is beneficial to others by warning them of the danger. The existence of the selfish gene in the role which Dawkins suggests would point to purpose behind the function of the genes, rather than the action being the consequence of blind chance factors. Let us therefore examine altruism in another light. Insects such as ants, bees etc. sacrifice their own lives for the good of the colony. The fact that these organisms are social and have evolved to live as a colony goes a long way to explain their behaviour. They share the same genes and the roles and function of the individuals are functionally an extension of the different cells of any multicellular organism which coordinates their effort to provide the organism with the opportunity to reproduce and pass on its genes. These cells too, when necessary, will sacrifice their lives in the defence of the organism in for example their combat against pathogens. In social insects the behaviour which served to protect the individual, also when having evolved to live in colonies, serves to protect the other members of the colony. The bird which produces a reflex response to danger by an alarm call also has the effect of warning other birds in the area. There is nothing here which would suggest that the function has arisen as a means of protecting the carriers of one's own genes. Should I be confronted by sudden danger I may respond by screaming, and this may well act as a warning to others of possible danger. It may also be just as likely to attract attention to the others close by, placing them also in danger, and therefore at the same time distracting attention from myself.

Human altruism, where people will risk their lives to save others, or even sacrifice their lives, is often quoted as an example of the selfish gene at work. This I contend is a poor example. Human behaviour is influenced by consciously taking on the values of society, where certain behaviour is rewarded while other behaviour is condemned. Bravery and cowardice are examples of how these two forms of behaviour are treated by society. Should I risk my life to save my son then I am attempting to preserve my genes. Should I risk my life to save a stranger I am placing my own genes at risk. It is hardly credible that this would play any part in influencing most behaviour, which is usually a combination of instinct and learning. Should one accept that altruism is an example of the way genes have evolved to protect themselves, and ensure their survival, then one must question how and why this ability has evolved for us to be able to control the future of our genes. From this one might suspect that genetic engineering has somehow evolved from the resulting evolution of our genes to provide us with the ability in the future to speed up the process of evolution. We are, I suggest, on very dangerous grounds. It is easy to fall into the trap of looking at the evidence and reaching the wrong conclusion.

A few years ago I owned a dog, and at different times of the day I would take him out for a walk. To my surprise, merely thinking of taking him out would result in him coming into the room and bringing his lead to me before I had even moved from my chair. This behaviour happened regularly and seemed to defy any explanation other than the dog having the ability to read my mind even when he wasn't in the same room. It was some weeks before I found the answer. Upon thinking of taking him out I removed my spectacles and folded them. The resulting click from doing this produced the associated response from the dog that we were going out. The simple explanation proved to be the answer, rather than the possibility of a telepathic ability which had evolved from the dog's ancestors when they lived in packs.

Chapter IX
CAN DARWINISM SURVIVE?

So far I have raised issues which in my mind question the validity of our present understanding of evolution. This does not mean that the present-day answers to these issues are in fact necessarily wrong. It may well be in fact that Darwinism and the present-day theory that has evolved from it are correct. One would however only be sure of this when the questions I have raised have been answered. Failure to do so will always leave a doubt and provide ammunition to support any new so-called religious belief. The present danger is that once an idea is sown and takes root, it can quickly become the basis for, and come to dominate, all further thinking. Darwinism was eventually accepted as an alternative to creationism because it could take its place. The fact that it will remain until such time that it can be replaced by an acceptable alternative theory does not mean that it is correct. The universal belief that the earth was flat, and later its central position in the universe, was eventually replaced, albeit with great resistance, with the beliefs we have today. It would seem that any theory which apparently reduces human importance will always be met with resistance. However at the present time the only alternative is to return to the biblical story of creation, and the fact that this has happened in certain circles supports the fact that Darwinism is not totally convincing. Galileo, and of course Darwin himself, found it difficult to change set ideas and beliefs. It is part of human nature which, both in its arrogance and in its deep-rooted fear of change, finds it unacceptable to admit that it has been wrong. Consequently it will battle on, changing and modifying a principle or an idea rather than acknowledging that the original concept was fundamentally wrong and should be abandoned if progress is to be made in the search for the truth. This is obviously easier to achieve when there is an alternative theory to take its place. In the case of Darwinism it would at the present time leave a void in our understanding of the world and our position in it.

Political ideas and social concepts become weapons to target the establishment and produce change. History is littered with accepted theories which once having been sold to the population are then difficult to abandon, even when they don't work and never will. Instead they will be modified and adapted in the hope that they will work, because it is difficult to admit that the original concept was flawed. Unlike blocks of sky-high flats, they cannot so easily be demolished and replaced. They become part of the social culture and as such are difficult to remove without creating a feeling of deprivation and mistrust.

Let me give two examples, which do not necessarily reflect any personal opinion. Social reform has gradually led to a reduction in the severity of penalties given for crime against society; the abolition of the death penalty, hard labour for convicted criminals etc. The theory is that this type of negative approach to crime leads to increased criminal behaviour, and that these social deviants would respond more to understanding and reform. The result has not been the expected reduction in crime, but an escalation at all levels. Instead of now admitting that the theory was wrong, a whole complicated pattern of reasons are given to explain the rise in crime. Unemployment, poor living conditions etc. are quoted, none of which have any provable validity, but instead serve to distract attention from the simple fact that a reduction of punishment does in fact lead to an increase in crime. It is of some interest that the penalties given to the convicted Great Train Robbers exceeded that for murder. Presumably this was intended to deter any repeat of such robberies. It seems to have worked.

My second example is in the field of education. Comprehensive education for all was introduced into England in the 1950s. All pupils were to be educated in the same way, regardless of the simple fact that all children have different abilities and intelligence. The results were that standards in education fell to a record low. The remedy followed the same standard response. Alter and modify to try and make it work. The matriculation system was abandoned and replaced by a steady number of progressively watered-down examinations, until at the present time, after the introduction of a whole range of grades, it is extremely difficult to fail. University entrance levels have had to be lowered, easier degree courses introduced, and industry has sometimes had to set its own exams to try and establish the educational levels of its future employees. Schools suffer from increases in truancy and disruptive behaviour, and this has led to teacher shortages. Yet still no one actually admits that the original idea was wrong.

Is science any different? As I write, evidence which supports the Big Bang theory of the origin of the universe has been placed in serious doubt. After the Big Bang the material expanded and so the universe as we know it was created. It has been shown that it is still expanding but will slow down, and will then either remain static or start to contract. It is quite obvious that matter in the outer areas of the universe furthest away from the Big Bang will be travelling slower. Unfortunately for the theory, recent observations have shown that matter on the outer edge of the universe is in fact travelling faster, which is the opposite to what should be occurring. Will this information affect the theory? I am fairly confident that it will be ignored. Who would be brave enough to say we were wrong without being able to explain why?

Should scientists always tell the truth? In the past it was dangerous to do so. The main difficulty and adversity faced by Darwin was his belief that man had descended from the apes. In people's minds man is different from other life forms and therefore must be separated from the rest of the animal kingdom. It took a great deal of indoctrination to change people's views. A belief in our superiority over other animals is akin to feelings of superiority over other races with all the possible conflict to which that path could lead. The concept of equality at all levels became the way to control the population, a backlash against the equally extreme social concept of the survival of the fittest. Equality may not always be believed by the individual but collectively public opinion has been conditioned to defend it against any suggestion which challenges this concept. An American scientist conducted a series of IQ tests on children and published his findings. His results showed that coloured American children had on the whole a significantly lower IQ than white American children. The condemnation of his findings were so high that he needed police protection. Racial prejudice can be dangerous in two ways, both arising from factual ignorance. The first is the belief that all men are equal, instead of the common sense view that all men are different. The second one is that a social truth is only acceptable if it places the recipient in a good light. It has been suggested that the testing was flawed and while the results may or may not be true, had the scientist shown that coloured Americans made better athletes than their white counterparts, or had a greater aptitude for music and rhythm, then this would have been readily acceptable. The point I am making is that only by being impartial to one's prejudices and applying objective thinking can science progress in its search for the truth.

People are of course conditioned in the ways in which they perceive and accept the world about them from the culture of the society into which they are born. It is impossible for the developing mind not to be influenced by these ideas. The basic questions of the individual of 'where do I come from?' and 'where do I go when I die?' are answered by either the creationist's belief in the story of Genesis or by the believer in Darwinism. The argument is that only one of these concepts can be correct. There is of course the equally valid view that they may both be wrong. This is far less acceptable. People are conditioned from birth to have answers and explanations for everything. It would be rather refreshing if some politicians were to state that they had no idea why, for example, pregnancy among schoolgirls was increasing at such an alarming rate, rather than stating that it is due to lack of sex education, availability of pornography etc., with no evidence beyond his or her belief that this is true. Again the point I am making is that we demand answers from the experts, and the experts in turn feel obliged in order to justify their position in society to supply them. When a person visits his doctor and describes to him the symptoms from which he is suffering, he expects the doctor to know what is wrong. Should the doctor say, 'I haven't the slightest idea what is wrong with you,' the patient would leave feeling very let down. Doctors, I suspect, know this and the ambiguous 'virus infection' has become a much more acceptable answer. The common sense reality that a doctor having taken your pulse and sounded your heart cannot somehow identify something as small as a virus is completely ignored. The question has been answered and the patient is satisfied.

In one's demand for answers the response can be interpreted to suit the one you would prefer. In the BSE crisis in Britain in the 1990s the politicians asked the scientists if there was any evidence to show that the disease could be transferred across species, thus endangering humans. The answer, that they had at that time no evidence, was translated into assurance to the public that beef consumption was safe. The fact that the answer also implied that there was at that time no evidence to show that it could not be passed on to humans was ignored.

I have tried to show how set beliefs, once they become established, are difficult to change. Common sense and education do little to eradicate them. Religious concepts, astrological beliefs etc. become fixed into a society's culture and are clung to rather than face the uncertainty of life without them. Politicians feel obliged to fulfil their elected function and produce new policies which are inflicted upon the electorate with promises that things will be better as a result of the change. When did any newly elected democratic government say that any aspect of society was working well and should be left alone? As a result of this they cannot then of course admit to making mistakes or getting it wrong without losing face.

Scientists are no different, and new scientific principles are gradually infiltrated into a society which generally accepts them without too much thought. Overpopulation is dealt with by the simultaneous introduction of birth control methods, abortions, and the contraceptive pill, and the counter-measures to these of fertility drugs and test tube babies. This says something for the blind confidence the population has in science's ability to solve problems. Should scientists always tell the truth? The answer is that whether they should or should not, the fact is that they frequently do not do so. The excuse is that only when it is in the public interest should the true facts be made available. One could be critical of this approach, but the public do like specific answers and reassurance rather than factual evidence which although honest is unacceptable.

Is the present theory of evolution correct? The answer has to be that at the present time it has not been conclusively proved. There are too many unanswered questions. The laws of the universe upon which our understanding of the universe is based must have existed before the creation of the universe. The origin of life and its evolution is not separate from this. It is as much a part of it as is the formation of a star or the production of a galaxy. How could all this be possible? We are in the process of assembling a jigsaw puzzle without any understanding of how it was created and for what purpose. The evolution of life from simple life forms to complex ones like ourselves must follow the same chemical and physical laws which existed before life began. This evolution must be responsible for us to be aware of this fact. How can this be reconciled with the survival of the fittest? What evolutionary survival advantage is there in having the desire to understand the origin of life? It is difficult to believe that, as evolutionary theory suggests, our arrival was the result of chance. Had the dinosaurs not become extinct then the mammals would not have evolved, and therefore neither would humans. It is the stages of human evolution which are important and which pose the greatest challenge to evolutionary theory. The human brain's creativity and appreciation, and its inquiring nature, are unique. It is difficult to see how the human mind can have evolved as a result of chance. Why did Shakespeare write his plays? Why do artists feel the need to paint? Why do composers have the ability to produce music? Where is the survival advantage of these actions? More importantly, why does the human mind find them so satisfying? Can all this be explained by the production of certain proteins by the DNA? These questions are as important in the study of evolution as in trying to explain how one organ was able to evolve into another. One other problem we face is that at every stage in our history we are only able to arrive at the truth as we then believe it to be from our perception at that particular point in time. Historically we know that what were in the past perfectly reasonable explanations have proved to be completely wrong. I have an old book of natural history which states that eels arise from horse hair. Eels had been shown not to breed in captivity in ponds or aquariums. They had however been discovered in horse troughs in the morning when they had not been there the night before. It seemed therefore the only reasonable explanation, and much more satisfying then to say we don't know how they breed. Swallows in the same book are said to spend the winter at the bottom of ponds and rivers, hibernating in the mud, to emerge again in the spring. Migration was not yet known about, and seeing swallows skimming the surface of the water in search of insects and then vanishing for six months, it seemed to be the only possible explanation. Will our present theories in two hundred years time seem just as absurd? Well, we don't have to wait that long. I have a university textbook published in 1970 which gives the origin of our moon as part of the earth thrown off during the cooling process, resulting in the formation of the area on earth now occupied by the Pacific Ocean. Well, we did not have to wait long for that myth to be demolished. In the theory of evolution we may have to wait a little longer. Only time will show if Darwinism can survive, and for how long.

Have I been unfair in my criticism of the present-day approach to evolution? I think not! Other branches of science have allowed themselves the freedom of attempting to explain the concept of the universe with a virtually unlimited number of propositions. With the aid of mathematics they consider the possible existence of up to fifteen different dimensions, and the role of black holes, which may well allow one to move from one universe to another. They consider the existence of other life forms so advanced as to be capable of travelling at the speed of light. These are concepts found not only in children's comics but in university textbooks. However when we come to consider life on earth and its origin and evolution then it would appear that such freedom of thought must be discouraged. This I contend is ridiculous. Life is as much a part of and product of the universe as is a black hole. We are created from the products of the stars and this cannot be ignored when we attempt to produce an explanation for our existence. For biology to advance it must be prepared to remove the restraints imposed by the limits of what we are led to believe to be possible and what may in fact be the truth. The difficulty we have to overcome is for example that had we evolved without the ability to perceive colour, then colour would still exist but we would be unaware of its existence. In the quest for the understanding of life we must be prepared to remove such restraints imposed by our senses, and allow the conscious product of our brains to accept and explore the phenomena that may well exist even if we cannot at present prove their existence. Once we attempt to explain the origin and nature of life within the constraints of a three-dimensional universe, then we will run into difficulties. The origin and future of our existence will remain as great a mystery as does the origin of the universe itself.

I have mentioned how the earth seems so well prepared to support life and has all the necessary components required for the advancements of intelligent life forms. Fossilised fuels, minerals, water and radioactive materials, all essential for technological progress. Compare this with the rest of the planets in our solar system. In his comments on man's comprehension of the solar system, Lewis Thomas compares our own earth with the other disappointing planets in our solar system.

'The overwhelming astonishment, the queerest structure we know about so far in the whole universe, the greatest of all cosmological scientific puzzles, confounding all our efforts to comprehend it, is the earth. We are only now beginning to appreciate how strange and splendid it is, how it catches the breath, the loveliest object afloat around the sun, enclosed in its own blue bubble of atmosphere, manufacturing and breathing its own oxygen, fixing its own nitrogen from the air into its own soil, generating its own weather at the surface of its rain forests, constructing its own carapace from living parts, chalk cliffs, coral reefs, old fossils from earlier forms of life now covered by layers of new life meshed together around the globe. Troy upon Troy. Seen from the right distance, from the corner of the eye of an extra-terrestrial visitor, it must surely seem a single creature, clinging to the round warm stone, turning in the sun.'

Chapter X
PURPOSE?

Although it may be thought by some that I have been over-critical of the progress which has been made in the attempts of science to provide answers to the questions which arise from the study of living organisms; then I can assure you that this was never my intention. In fact much of the progress in our understanding of the natural world reminds me of the philosophy of Sherlock Holmes: eliminate the impossible and what is left, however improbable, must be the truth. The problem is that it is often very difficult to determine what is possible and what is impossible. The difficulty of attempting to understand how life arose, and then evolved to produce human beings, raises the most difficult questions that can be posed. To suggest that the answer is obvious and that we merely evolved from the apes, is both naive and arrogant. Historically we know that science has progressed through a whole catalogue of wrong assumptions, each of which at the time seemed beyond any question or doubt to be correct. The answers provided by religions have the advantage of removing the problem of having to provide any proof, relying as they do upon faith rather than factual evidence. It is easy to be critical of this approach, but at least by doing so they have recognised the problem of placing humans in a separate category to the rest of life on this planet. While there is no doubting that we have a great deal in common with the rest of the animal kingdom both in our physical structure and our ability to survive and reproduce; nevertheless there remains the tantalising difference between humans and the rest of the animal kingdom. The principle of evolution from simple to complex life forms appears to be correct, and the only real criticism is in the way this was achieved. Following the ways which I have examined it could not have resulted in the production of modern man since it defies the basic rule of evolution with its rule of the survival of the fittest. Humans have evolved a mind and behaviour which is not in keeping with Darwinian evolutionary principles.

The early anatomists spent a great deal of futile effort searching for the human soul. Although conscious of its existence they were unable to find it. It was rather like a child trying to find the people inside a television set; they can be seen and heard but are not there. Humans have compassion, sadness and joy, and an appreciation of beauty. We show anger at injustice, and love, which is difficult to define. We fight wars for what we believe to be just causes, when both sides consider the other side as the enemy. We are conditioned to know right from wrong and to accept the rules of the societies into which we are born. We accept the inevitability of death with or without the promise of eternal life, and not let it interfere with our concern for a future in which we will play no part. All this and more, if we accept present scientific ideas, evolved without any purpose, but just by chance.

Keith Ward, in his book *God, Chance and Necessity*, believes that the function of complex bodies is to make possible the development of the central nervous system and the brain which can receive information from the environment and respond to the information. Bodies, he states, are not primarily machines for carrying genes. The purpose of the bodies is to build brains whose purpose is to build consciousness and purpose.

It would seem that to remove purpose from the equation is just plain silly, and the only justification for doing so is that it removes the problem of trying to explain something which at the moment we are incapable of doing. There is a growing tendency to ignore anything which knocks at the door of so-called scientific facts. In the 1970s or 1980s, a documentary programme was shown on BBC television. It was a programme of great significance but unlike other programmes was never repeated. It was made in a London hospital and was produced as a result of the events which followed on from what is unfortunately a not uncommon occurrence. A man in his late twenties was admitted to the hospital. He had suffered severe head injuries resulting from a motorcycle accident, and was unconscious. A routine scan of the skull was carried out to assess the damage. The scan showed that the skull cavity was empty of brain tissue. It is a rare condition, when the skull is filled with fluid before birth and normal development of the brain is prevented. A primitive nervous system around the inside of the skull allows the person to live but any intellectual development is impossible. After birth the child will live but in a physical and mentally handicapped state. Upon contacting the parents of the man, the doctors were surprised to find that they had not been informed of his condition at birth and he had led a perfectly normal life in both his education at school and later at work.

The hospital decided to investigate. Maternity hospitals were contacted throughout Europe and a follow-up contact was made with other babies who had been born with the same defect. Six to eight people who had been contacted agreed to come to the hospital and have the scan. In each case the scan showed an absence of any functional brain. They were all mature adults and all had lived perfectly normal lives, being completely unaware of their condition. One had led an academic career and had a teaching post at a university. None of the people examined had ever been aware of their condition. The neurosurgeon who had conducted the survey and investigation admitted that the result made a mockery of his specialist knowledge, but he had no choice but to continue and disregard the findings, which were totally unexplainable. As far as I am aware the matter was closed. I apologise that I am unable to give names or dates. The programme was not repeated and I have never seen any mention of it in any book or scientific journal. Censorship? I will leave it for you to decide.

This may well have proved to be one of the most significant findings in the attempt to understand the truth of the purpose behind our existence, and it has been ignored. There are others of perhaps less importance but which do nevertheless pose a threat to our present beliefs. I have already shown the difficulty of attempting to explain the way behaviour is passed on from one generation to another when the behaviour can only have resulted from having been learnt by experience. The answer we are given for this is that certain genes cause certain behaviour to occur which if advantageous to the individual will be passed on and so gradually increase in efficiency to become complicated behaviour such as nest-building or web-making by spiders. This seems to be an unsatisfactory explanation. In the case of web-making the spider would have to have evolved organs in order to manufacture the silk, and spinnerets to weave a web before the ability to produce a web had evolved. To put it simply, it would be no use someone providing me with a whole set of carpenter's tools if I lacked any knowledge of how to use them or indeed what they were for. They would certainly give me no advantage over my fellow men until I had acquired the skill to use them, and this would have to be learnt. The spider was even more handicapped; it had to wait and hope that a series of chance mutations would somehow produce the correct behaviour in order to be able to use the tools. In order to demonstrate how evolution depends upon the survival of the fittest, many rather strange and questionable examples are cited. Many insects which are poisonous if eaten are brightly coloured, which supposedly warns predators of the danger, and they thus avoid being eaten. How can birds learn this and pass this information on? Well, certainly not by trial and error. Other insects such as butterflies have supposedly evolved coloured wings which will provide them with a camouflage against predation from birds. Meadow brown butterflies are however hardly inconspicuous against grass, and logically green would be a more appropriate colour. The distinctive eyes on the wings of the peacock butterfly are supposedly there so a predator would attack the wings and not the body of the insect. Why should this have evolved, despite the fact that it makes the insect more conspicuous? It is interesting that the most common butterfly, the cabbage white, has no camouflage at all and is probably the most easily seen. The information I have quoted can be found in most butterfly books giving the reasons for the coloration of the wings. It just does not stand up to the application of common sense.

The geneticists state that the number of genes in a large organism such as man can be 100,000, and that mutations can result from a whole range of factors resulting in changes in the make-up of the genes. Random nucleotides and substitution in the genes can result in changes in anatomy, behaviour and physiology. Genes may also shift their position on the chromosomes, producing different phenotypes. If these changes increase favourable advantage then they will spread through the population, otherwise they will die out. On the face of it this sounds reasonable, but it has to take into account that the mutations must correlate to produce the total change which would be beneficial. There has to be coordination between the various sequences in order to produce a significant change.

In the higher order of insects the egg hatches into a larva, which gradually increases in size through successive instars and then goes through a pupal stage. Here radical transformation takes place, the adult structure develops by special cells, imaginal discs which are inactive in the larva, the wings actually developing internally from the body wall. How could this be the result of a sequence of mutations which correlate? The whole process can only work as a whole. Gradual change in the process would be useless. A child asks how did the caterpillar turn into that butterfly. 'Oh, it's all due to chance mutations,' says the evolutionist. 'I don't know,' says the child's father, 'it's a miracle.'

It is taken for granted by the evolutionists that the mutations occur by chance and ignore the fact that it is more likely that they are programmed to change and evolve. The eye according to Dawkins has actually evolved up to sixty times in many different invertebrates. Would this not suggest that this points to the following of a preconceived plan?

I contend that we cannot ignore any possibility until we can explain the meaning of the whole universe. Theories as to its origin and purpose shine like confetti in the unmoving dogma of evolution and creation. The reason is not difficult to see. Darwinism plays no part in this research, thus allowing them the freedom of looking at any possibilities. The search for life elsewhere in the universe presents a great many problems. It is generally assumed that life would need the conditions which occur on earth. This is rather a strange assumption, considering that life on earth has the ability to thrive in conditions as extreme as it is possible to imagine. Some forms are not even dependent upon the sun as a source of energy for their survival. Equally evolution on earth has been dictated by the prevailing conditions which it had to overcome. Since these conditions will be different elsewhere in the universe then the life forms may well be difficult to recognise. Certainly it would seem that life evolved on earth very early on and did not have to wait until conditions had settled down to suit the types of life which we find today. It is even possible that we would not even be aware if life did actually exist on other planets. All the methods we use to identify life forms have evolved in keeping with our ability to detect life on earth using our senses which has evolved for that purpose only. It is not difficult to see the problems that awaits future scientists in trying to discover if we are indeed alone.

EPILOGUE

I, along with the millions of other humans who exist today, started my existence from a tiny egg that had by chance united with a single sperm. Just as a card player on being dealt his hand knows that chance has decided the value and worth of each card, and it will be up to him how the hand will be played, so as the fertilised egg divides and grows, there develops the individual's ability and choice to decide how each genetic card will be played. High-value cards could be used to maximise their returns or be thrown away with complete disregard of their value. Compared to other life forms, and because we are human, we are given the privilege of playing this game of life with the ability to choose how we play our hand. Somewhere in the genetic pack we receive an extra card. It has no face value, only the word CHOICE. It is the most difficult and demanding card to play, placing a responsibility on humans which has not been given to any other life form. It allows us to shape our own destiny, but more than that it allows us to help shape the destiny of other players in other games. We the players are the only ones who can decide how the game progresses. The other life forms play their part but have no more influence on the outcome than has the life forms that produced the raw materials from which the cards are made. This special card named CHOICE allows us to look at the evidence and decide what is true and what is false.

All but the simple animal forms communicate, but only humans have the words 'how', 'why' and 'what' in their vocabulary. It is these three words which manifest themselves and are present in children at an early age. They are not acquired by them, but are as much a part of their natural development, as is the building of their muscles or the hardening of their bones. It is society's ability to try and answer these questions which will allow them to form opinions of their own and pass these on to the next generation. These opinions taken on board at an early age can become fixed and difficult to change, becoming naturally resistant to new ideas. Darwinism faced this difficulty when challenging religious beliefs. Having finally received a general acceptance by most people it must still be allowed to be questioned, not sheltered from too close a scrutiny.

I have examined the two alternative explanations for the production and evolution of life on earth. Religions formed the basis for providing the answers to this problem, and they gained strength and support because they were at the time the only answers available to account for human existence. The bonus religion had was that it provided a moral code, which in turn provided a guidance for the freedom of human behaviour. Darwinism and science generally may well challenge the concept of God as a creator, but religions will continue to survive because of their usefulness as a political control of free will. Did not Marx consider religion as the opium of the people? Nevertheless religion based upon blind faith and its support for the importance of man has stood up well to the challenge of science. It is not difficult to see why. To categorise man as an organism which has evolved by complete chance and without purpose provides little comfort to a creature that is born to be aware of its own destiny. Life without hope would be a purposeless exercise and societies would break down. I do not believe this will ever happen, at least not until man has fulfilled the purpose for his existence. I find it interesting that the great champion of Darwinian evolution, Richard Dawkins, while refusing to accept any purpose behind man's evolution, accepts that consciousness is 'the most profound mystery facing modern biology.'

It was as a direct result of the inquiring mind of man that led to his religious beliefs, and now science faces the same challenge. People seek answers, and with the great advancements in communication they are able to receive them. The media rises to this challenge, but just as Sunday schools were used to water down religious concepts for children, so science is watered down by the media. Television provides popular science using every trick in the book. Evolution is depicted with the use of 'live' animals, dinosaurs in colour, complete with sound effects, and the public accept what they see as being true. Entertainment – yes, education – no. Science is sold to the public in the same way as an estate agent sells property. He emphasises the good points but avoids mentioning the flaws. The Darwinian theory and religious beliefs both have valid arguments, but both are guilty of ignoring the flaws in their make-up. The public are now being sold the 'gene theory', whereby just about everything can be explained. I accept that the physical make-up of my body is built to the design of my genes and may well be beyond my control, but unlike other life forms we have the ability to choose how we live and what values we develop. We have minds which from an early age demand answers. They should not be fobbed off with half-truths. Stephen Hawking's book *A Brief History of Time* became a bestseller, although it is doubtful if any but a handful of readers managed to get beyond the first chapter. It demonstrated however that people hoped to find answers to questions that went beyond what was needed for their survival.

The universe was created by the Big Bang. No, these are not the words of the theologians but the words of the cosmologists. What is interesting is the word created. Included in this word created we have to accept that all the laws of science which over the years of human enlightenment have allowed our knowledge of the natural world to advance, were also created. It can, I feel, with some validity be argued that these laws must in fact have existed before the Big Bang. Nevertheless these laws dictated the way atoms behaved to produce the elements and the means by which these combine to produce the unlimited numbers of compounds which are the domain of the chemists. The laws of mathematics came into existence, and as they were gradually discovered provided the tools for the mathematicians. The physical laws, such as gravity, enabled the formation of the galaxies, stars and planets. They also provided the ingredients for life. 'We came from star dust,' were the words of a cosmologist speaking to children at the Royal Society's Christmas lectures. How can science deny with any confidence the theologians' concept of a creator when it supplies so little evidence against such a concept, and when each new discovery denies any other explanation? Quantum physics has now opened up a whole new field in the laws which control matter at sub-atomic level, to place further strain on conventional science. Consider our present biological theories in light of what I have written. Why have we evolved to fathom out these matters when all that was required from evolution was for it to improve our competitive ability to survive in order to reproduce?

I have stated that even if one accepts the fact that life started by accident with the chance coming together of material that had the ability to replicate itself, then why, I ask, did this become the main driving force of every organism? As I have previously stated, all life exists for the sole purpose of reproducing, and then dies. The concept of the survival of the fittest is concerned with changes which give an individual a competitive edge in survival. Reproduction is not to the individual's advantage for its survival, only for the survival of the genes. This points to purpose and as such allows evolution to occur. To suggest that this just happens to occur by chance, and the resulting unexplained obsession by the genes to be passed on, resulting in the eventual production of intelligent humans, makes no sense.

I have looked at the evidence and must conclude that there is purpose to evolution and that it has still a long way to go. Is it not possible that not only are we made from star dust, but it is in the stars that our destiny lies? Man has shown already his ability to produce the technology which may well in the distant future replace him, just as today's advanced computers are replaced by those with greater capabilities. In the future they will no doubt evolve to be self-maintaining, then man, like all the life forms before, will have fulfilled his purpose in the programme of evolution, and become extinct.

Am I right? I don't know. I cannot see into the future. However if I am walking along a road on which I have never been before, then I may not know where it leads, but I do know I will eventually reach somewhere that was guaranteed by the person who made the road.

So far, the evidence that I have examined for the emergence of life and its gradual development from simple life forms to more complex ones, and from these to the eventual arrival of the present human population of today, does not fit the general explanation provided by conventional science. Every step points to the concept of purpose and intention. One can appreciate the attractiveness of explaining the whole evolutionary process as a series of chance events which were under the controlling effect of the Darwinian rules. These rules, we are led to believe, were applied once life had started, and without any purpose just happened to result in you and I being here at this point in time. The origin of life itself presents a problem, and this cannot be fitted into any form of convincing explanation as to how or why it should have occurred. Again the chance factor becomes a more attractive alternative to the theory that it was the beginning of a set plan that existed even before the whole evolutionary process started. The difficulty faced by the present evolutionary theory is of course man himself. The present scientific champions of the modified Darwinian theory face the insurmountable and therefore generally ignored problem of explaining why they themselves have evolved to find it necessary to seek an explanation for their own existence. Unlike all other life forms, man alone attempts to discover answers to his questions as to his existence and purpose. Religions arose before science and will continue to exist as long as science remains impotent in providing a satisfactory answer to their questions, which are as much the result of evolution as are the brains from which they originate.

Unlike the evolution of their brains, from a Darwinian view they provide no reason for their existence since they provide no advantage for survival. On the contrary, time and effort spent in the pursuit of attempting to answer these questions should be a handicap in diverting time and energy away from competition and survival.

Taking the driving principle of evolution as being the survival of the fittest, then man has evolved to generally disregard this principle. The weaker members of society are generally excluded from the normal competition for survival. Physical and mental disabilities which would reduce the survival competitiveness of such individuals tend to be ignored. Humans have evolved a conscience which allows the natural laws of evolution to be disregarded. The reason for this may well be religious influence and the free will of man to overcome the natural instinctive behaviour of gaining personal survival advantage. One can argue that individual ability and intelligence places every individual with a competitive advantage or disadvantage and that this in turn will influence his ability to provide for his offspring. This would be in keeping with the principle of the survival of the fittest. Nevertheless in an advanced technological world the principle that all men are equal must influence the evolutionary effect of this principle. Medical resources are equally shared out, or are at least under moral and political pressure to do so, with no consideration for the return benefit of such behaviour. The old and infirm are cared for with no thought of the drain upon the resources of the stronger and more evolutionary valuable members of the human race. It may be that religious belief which gained strength from science's inability to explain the role and ultimate destiny of man, has backfired by throwing a spanner in the works of the evolutionary laws. Equally it may be that evolution with the arrival of modern man has no longer to follow the principle of the survival of the fittest. Man's arrival on earth has proved to be a turning point in what had occurred before. Does this mean that evolution has reached its climax? I think not. The earth is not static. It is easy to believe that it is tailor-made to support life, and to some extent that is true, but more importantly it is designed by its periodic dramatic changes in climate to provide the stimulus for evolution to occur. Life has generally adapted to these changes and the species that were unable to adapt became extinct. What then of modern man? Primitive man was able to overcome the dramatic changes in the world's climate, but modern man faces the same rules of evolution. The more specialised an organism becomes the less adaptive it is to change, and extinction becomes a greater possibility. The question one faces is, will present-day humans be able to survive in the future?

Climate is one factor which dramatically demonstrates its effect upon the human race. The more advanced we become, the more vulnerable we are to relatively small changes which create devastating effects upon the human population. Imagine the devastation which will be created by the next ice age. It will be a long winter of discontent. Domestic animals artificially bred will no longer have the ability to adapt to dramatic change in the environment. Selective breeding has seen to that! Modern farming methods to satisfy the increasing demand for food have become specialised and intensified, making farming more vulnerable not only to disease but to climate change.

Climatic change will also increase demand for the supply of energy from the raw materials of oil, fossilised fuels and the production of electricity. Supplies of these will diminish and this must result in a reduction of the human population as demand exceeds supply. Even without this the supply of raw materials from the earth's bank of minerals is at present being used up at an increasing rate. The conservation of these resources can only be achieved by a general population reduction. Modern society is geared to their use as though there were no tomorrow. A reduction in population would reduce the demand before the supplies are used up. From a resource's point of view the earth can be compared with a great spaceship which was fitted out with all the essential materials needed for its crew on their journey. Once the journey had started the spaceship became a closed system, with no further input. The longer the journey the more likely the supplies will run out. The realisation of this factor has finally got through and resulted in rather weak ideas for controlling the indiscriminate use of the world's natural resources. The indiscriminate use of copper and silver has been confined to essential uses and the recycling of used materials is beginning to be taken seriously. Too little, too late, perhaps. The point is that if evolution is to continue then the rapid increase and turnover of the human population which was necessary to produce the few individuals who have contributed to the present state of technological advancement and who will continue to do so in the short term, has to be reduced. This, I contend, has already started and will continue to do so. There are already signs that male fertility has significantly reduced without any convincing medical reason. Pathogenic diseases, although initially controlled, are beginning to become a serious threat. They are able to mutate at a far greater rate then the scientists' ability to produce the methods to combat their effects. New counter-methods to fight their potency are rapidly running out. Superbugs are a reality, and the full devastating effect they will have in the future upon the human population is a certainty.

I have painted a picture of the future which I believe is inevitable. The more specialised a species becomes, the more vulnerable it is to extinction. Humans evolved the ability to advance way beyond that which was necessary for individual survival, and this I believe wasn't the result of chance. It was to pave the way for the next stage of evolution. Humans, however, still retain the basic patterns of behaviour which are inbuilt into all advanced life forms. We live in houses and are happy to live in communities, but this does not prevent us from defining our territories with fences and hedges. We respond to a need to defend our territory, be it an individual's property or his country. When individuals or groups feel threatened and are unable to respond in this way, then the result is anarchy. This phenomenon is noted in the study of animals in zoos, particularly among the primates. In man the moral codes and rules of society are broken down. Family life and its role and responsibility declines and there is no way back. In the future the human population, rather like the dinosaurs, will become virtually extinct. What will take its place? The dinosaurs had the mammals waiting in the wings. We have technology.

The next stage in evolution will be machines, self-replicating and immune to the rules of survival. This I feel will be the next stage, although I hesitate to say the final stage. This I believe will be the future destiny of evolution and the aim of all that has gone before. This was the only way it could be achieved, from the first coming together of a self-replicating molecule, to the production of a self-replicating intelligence. Should I be wrong then the earth will end up as it began. A planet full of future promise awaiting for a miracle to happen.

BIBLIOGRAPHY

Bateson, P., and Martin, P., *Design for Life*, London, Vintage, 1999
Brown, A., *The Darwin Wars*, London, Simon and Schuster, 1999
Dawkins, R., *Climbing Mount Improbable*, London, Viking, 1996
Davis, P., *The Fifth Miracle*, London, Alan Lane Penguin Press, 1998
Denton, M., *Evolution, A Theory in Crisis*, London, Burnett Books, 1985
Eldridge, N., *Reinventing Darwin: The Great Evolutionary Debate*, London, Niles Weideufeld and Nicolson, 1995
Gallant, R., *How Life Began*, New York, Four Winds Press, 1975
Gribbin, J., *Genesis*, London, Dent, 1981
Hardy, A., *Darwin and the Spirit of Man*, London, Collins, 1984
Hooper, J., *Moths and Men*, London, Fourth Estate, 2002
Judson, H. F., *The Eighth Day of Creation*, Cape, 1970
Maddox, J., *What Remains to be Discovered*, London, Macmillan, 1998
Malik, K., *Man, Beast and Zombie*, London, Weideufeld and Nicolson, 2000
Milton, R., *The Facts of Life*, London, Fourth Estate, 1992
Murphy, E., *Beyond Darwin*, London, The Book Guild Ltd., 1995
Sheldrake, R., *The Presence of the Past*, London, Collins, 1988
Silvers, B., *The Ascent of Science*, Oxford, Oxford University Press, 1998
Rattray, Taylor G., *The Great Evolutionary Mystery*, London, Warburg, 1983
Thomas, L., *Late Night Thoughts*, Oxford, Oxford University Press, 1984
Tudge, C., *The Day Before Yesterday*, London, Cape, 1995
Ward, K., *God, Chance and Necessity*, Oxford, One World, 1996
Wilson, E., *In Search of Nature*, London, Penguin Press, 1996

www.ingramcontent.com/pod-product-compliance
Lightning Source LLC
Chambersburg PA
CBHW020444220526
45464CB00002B/844